Pete and Maggie were both born in Scotland and still live in Ayrshire. Pete is an authentic hippie born in 1955 and Maggie, born five years later, was a child of the sixties.

They met at a local library where Maggie worked for 15 years and were good friends for many years before eventually getting together as a couple.

In loving memory of Elizabeth 1976–2021

Pete and Maggie Burleigh

THERE AND BACK AGAIN

AUSTIN MACAULEY PUBLISHERS™

LONDON * CAMBRIDGE * NEW YORK * SHARJAH

A CIP catalogue record for this title is available from the British Library.

ISBN 9781398455054 (Paperback)
ISBN 9781398455061 (ePub e-book)

www.austinmacauley.co.uk

First Published 2024
Austin Macauley Publishers Ltd®
1 Canada Square
Canary Wharf
London
E14 5AA

We would like to thank everyone we met on our travels. We have changed names to protect anonymity, but you will always have a special place in our hearts and we will never forget you.

Many thanks to everyone at Austin Macauley who helped us get our book published.

Last, but not least, to Billy Connolly and his big banana feet, who seemed to be with us on our travels and has been making us laugh since the seventies. You are our Superhero Big Yin. Lots of love from two, eternally grateful, wee yins.

Table of Contents

Made Fae Girders

PETE:

In the beginning, there was light and we saw the light and we gave up our little cottage in the heart of the Ayrshire countryside in Scotland. To me it seemed like the middle of nowhere, yet it was not too far from civilisation. We asked Margaret's daughter to paint our hut with a hippie campervan on it, complete with a rainbow, peace signs, flowers and all sorts of groovy things. We Christened it 'The Peace Hut'. We had a fire outside most nights. It was so peaceful and quiet. Why would anybody want to give it up? One word: FREEDOM.

When we told folk what we were going to do, most of our family and friends thought we were 'nuts'. I don't know if our ages of 63 and 58 had anything to do with it, but to us they are just numbers. Well, maybe the campervan on the side of the Peace Hut was a hint (a giant Vision Board) because we love, repeat LOVE the great outdoors and the freedom it brings. We already had a very small campervan called Sheila. We went to Arran for a few days and extended it to a few weeks including visiting Mull and as far the Cairngorms. When we eventually came back to the cottage, our minds were already made up.

The freedom to move around and not be stuck in the same place. This was the life for us!

Our families are grown up and there was nothing really to stop us. We sat down to talk about it. When I picked up an old book, a map of Australia fell out. I had already found a brand new wallet on the ground a few days previous with nothing in it, but on the front was an Aborigine in one-leg stance, spear in hand, as if guarding a map of Australia… we both agreed it was a sign. It had to be Australia.

We also decided there is no point going to the other side of the world without visiting New Zealand as well and we might as well make it eight weeks in each country.

We had bought an old taxi which had belonged to a trusty friend and had it converted. We called him Gabie after my friend, Gabriel, who had recently passed away. We later found out that Gabie, who was a taxi driver, had agreed to buy this very taxi just days before he passed away. These cosmic signs followed us our whole trip.

The departure date was 28th January. We gave up the cottage on 1st October 2018, moved into our van and 'ate berries' until the day arrived and it was time to go. What we saved in paying rent we could put towards our travels. Australia, here we come!

Maggie:

When I was twenty-nine years of age and the 'BIG 30' was looming, I made some big life changes. Firstly, I learned to drive and passed my test. Then, I gave up my mundane secretarial job to go back to school and sit some Highers and Advanced Highers in the hope I would find a more fulfilling job. It worked. I expanded my mind by studying Renaissance History, English, Modern Studies and Biology and eventually ended up with a job I adored in a Community Library.

I suppose what seemed like big life changes then seem small compared to the one I made nearly thirty years later. Yet it felt like the most natural thing in the world to give up the constraints of a frequently frantic 'normal' existence and swap our house for a campervan. It gave us freedom to live an authentic spiritual existence: to just be in the flow with life, trusting The Great Mystery of the Universe.

It certainly raised some eyebrows, some horrified looks and some pitying comments from a lot of outsiders looking in. There were others who totally 'got' it. We knew in our hearts that we don't need a lot of material things to be happy. I had some Icelandic wool covers so we knew we would never be cold in bed. We moved into Gabie our campervan on 1 October 2018. We brought in the New Year in Aviemore in the Cairngorms. Pete kept finding Australian 'mojos' so we followed our intuition and here we are telling you of our adventures.

There's Something in the Air...

PETE:

After the beginning came the word and the word came in the form of the diaries we both decided to keep. Someone had mentioned to us why not keep a journal or even write a book. It might be interesting. Good idea, and by keeping two diaries, we could have both perspectives. So we parked our van, left our few possessions in storage, grabbed our diaries and the word became flesh.

Off we went to Glasgow Airport for our flight to Dubai, our first stop. All the stewardesses and staff had those lovely smiles that comes with the job. The girl we spoke to told us that double decker planes had been introduced. I would have thought she was winding me up if I hadn't seen double decker trains with my own eyes in Switzerland, so I believed her. The flight was long, but very comfy. We had scoffed our crisps before take-off, but all was well as we found out food was being served as soon as we took off. This was the start of regular food, snacks and free drinks; beer, wine, liqueurs and of course non-stop films. I watched something about wolves and Margaret watched Bambi or something.

MAGGIE:

The Emirates stewardesses looked very exotic with beautifully coiffured hair, veil like headdresses and their red lipstick. The lovely stewardess, who checked us in, smiled warmly at us. When she spoke we didn't expect a Glasgow accent, so, we were pure delighted when she had one. She was such a helpful lassie. So glad we arrived at the Airport in plenty of time for the flight as there was a wee blip re our visas. It was a bit of a hold the breath moment. However, after she made some phone calls, all was well.

The first leg of our journey began. I felt very pampered and also very excited, having only flown on budget flights previously. Pete watched something about wolves and I watched Green Book which is based on a true story, had won Oscars and stars Viggo Mortensen. Viggo, when playing Aragorn in Lord of the Rings, is my ideal man, apart from Pete of course. He also happens to be a great actor as well and it's a marvellous film. Was so happy to see they also had loads of Disney films. I love Disney. So much wisdom for young and the young at heart. Decided on the Lion King (old version). Can you feel the love tonight?

PETE:

We landed at Dubai and stretched our legs, passed police with guns and were a bit afraid to buy anything in case they were ridiculous prices. Mind you, can't get any more ridiculous then paying £5 for a packet of M&Ms at Glasgow Airport. We decided not to bother.

Back on the plane, it was a different plane, not a double decker. It was the fanciest plane I ever saw, with reclining leather seats, mini bar attached, loads of leg room and fancy TV. However, it only took a couple of minutes until we were told to move on and were led into a different section. No leather seats but still comfy, still free drinks but, you have to ask for them. To be honest the only difference is the leg room. Everything else is basically the same. Then that was it, hours later we landed in Brisbane, Australia. Time to get the map out and decide where we are going first: Uluru, Darwin, Sydney? JINGS! we weren't expecting Australia to be THIS huge. So big that New South Wales is the size of France. We decided that Uluru would have to wait, so would Darwin and Perth and anything else that was not on the East coast.

MAGGIE:

Glad to be back on the plane. Couldn't be bothered walking around Dubai Airport at four am. I was asked to go in a wee room which slightly alarmed me. I then realised it was for my own privacy, as I was searched by a really pleasant lady and let go with a nod of approval. Back on the plane we had a couple of liqueurs with our coffee and I managed a good sleep.

As we neared Brisbane I got chatting to a young girl sitting beside me. She was a young exchange student journalist who lived in Brisbane. She had just spent a year in the UK doing a work placement. I asked her what she recommended we see in Brisbane and she said to visit Kangaroo Point at the south side.

When we got into Brisbane, it was 10.30 pm and we had left Dubai at 5 am. I will let you work that out because I am still trying to. I had booked an Airbnb room in the west end side of Brisbane and we luckily found an airport shuttle bus to drop us off. We were last off. The driver walked us to the apartment and waited to make sure we got in OK. I had noticed earlier that he was very kind and helpful to the other passengers. This really was the beginning of a beautiful friendship between us and the Aussies.

Possums, Priscilla and Other Amazing Creatures

MAGGIE:

The first thing I noticed about the apartment was an Aboriginal painting of a whale. A thing of beauty. It was so cleverly done, as if looking down from above. Seemingly these paintings are now highly desirable. I was looking forward to finding out more about the real Aboriginal culture as we travel.

Our bedroom had a magnificent view of the Brisbane skyline all lit up, so we kept the blinds open all night. Next day, despite jet lag, we went for a walk. The West End had a very arty, laid back vibe. I was surprised to note that there were lots of vegan and vegetarian restaurants. It was actually our hostess's dream to open a Vegan Tapas bar which seemed highly courageous to me. We found out later on our trip that Australians are actually some of the biggest meat eaters per capita in the world just behind Americans. Pete saw a possum in the park as we walked back to the apartment. I missed it. Certainly did not expect that in the middle of a city.

Caught the train to Gladstone. Saw our first kangaroo. Although I know what it looks like from pictures and TV, there is something surreal about seeing it for real. It is the craziest looking animal I have ever seen. Hopping up the field, it reminded me of Billy Connolly in his big slipper. This image stayed with me the whole journey which kept me entertained.

We are staying Airbnb another two nights before embarking on a trip to Heron Island to see the giant turtles and hopefully baby turtles. After that we will be off to get our campervan for the rest of the trip.

When we got off the train, we could hear a commotion in the distance. We were not sure what was causing it, thought it might be some sort of factory. It's late afternoon so we got a taxi to our Airbnb. Priscilla and her husband are our hosts and we are staying in a trailer tent in their garden with a raised up, double bed which ends up being super comfy. The thing that caught my eye when booking is that Priscilla makes a barista style coffee in the morning along with providing breakfast. Looking forward to that. Woke up early as we got an unexpected alarm call. Kookaburras doing their cackling laughter calls. I just had to smile. I feel they remind us humans, certainly me, not to take life so seriously. They also gave me a fond reminder of my ex-husband. He used to do a great impersonation of them.

PETE:

It was agreed that we stay in Brisbane a couple of nights then head to Gladstone so we could go to Heron Island to see the turtles. We went a walk the night before we left and I saw my first Australian wildlife—a possum—little did I know at the time that we would see lots of them and I mean lots.

We reached Gladstone after a six and half hour train journey which we enjoyed. I even saw my first kangaroo. It had huge flat feet and was bouncing over a piece of land so quick I was lucky to see it. I was a bit peeved it didn't have boxing gloves on. We got something to eat on the train, not in the same class as the aeroplane though. We were offered an assortment of pies. We were told by the trolley lady the Aussies like their pies, so we thought we better adapt to the Aussie way of life which was quite easy as Scots like their pies too.

We finally reached our first stop—Gladstone. It was quite late but we kept hearing all this noise. We weren't sure what it was. Got to where we were staying and eventually went to sleep. Wasn't that long until we woke up to more noise only this time it was kookaburras laughing their heads off at half five in the morning. Margaret fell in love with them immediately—it was to become her favourite bird—and they seemed to follow her everywhere we went.

Once the kookaburras had woken us up we had coffee with our host Priscilla. Great coffee as she used to have her own coffee place, kept the machine and some coffee. Turns out her husband and her are devout Christians. I'm more than fine with that, as at least you know that they are not going to mug you. In fact, they seemed happy folk and I found out they like nothing better than eating a fruit bat with a nice red. She said it was quite a dark meat and in fact it was on their honeymoon on one of Polynesian islands that they had it and it was delicious. I'll just take her word on that one and after a bat free breakfast we went a walk and came to the river walkway.

As we got closer to the trees along the banking I was quite sorry to see loads of black bin bags all over them. It had been so clean and litter free until then. Suddenly, one of the bags took off. They were, in fact, Fruit Bats. Hundreds of them. They had red, orangey chests and they were all actually hanging upside down. Now and then they would open their arms and close them again just like Christopher Lee in the Dracula Films. They were bigger than our bats, much bigger and I could see they would feed two folk quite easily.

Further along the way I saw a hose and a water tap. I ran straight up to it and stood with my head under it. Awwww man it was great. That's how warm it is in Australia.

MAGGIE:

Priscilla is really interesting. A lovely young woman. True to her word we had a real barista style coffee, made with all the appropriate noises. She used to have her own coffee shop before having her daughter and now there is another baby on the way.

She spoke about having a Doula, which is the name given to her birth adviser and who she can turn to for any advice she needs. She was talking about the conscious elimination potty training AKA elimination communication which she was doing with her wee daughter (we had met a Swedish couple last year doing the same thing). Found this very interesting and felt this was way more advanced than back home. Yet I find out that it is an ancient way of doing things practised particularly in African and Asian cultures. It takes time and patience and it relies on a very close relationship with parent and baby. It seems to be worthwhile for both baby and parent and takes place in the first six months of a baby's life. However because of people's lifestyles in the western world it is unlikely to achieve widespread uptake. To be honest I think my mum's generation did it and it's probably disposable nappies which have encouraged its demise for both these young mums, in a quest to save the planet, were not using disposables.

We enjoyed the long walk into town. Ten minutes into it, right next to us in the local park, we saw five large white cockatoos with very impressive yellow crests. The wildlife in this country is breath-taking and very, very exciting because you just don't know what you are going to see next…

I could not believe my eyes when we realised there were hundreds of Fruit Bats hanging on the trees in broad daylight. They are much bigger than I expected. They had foxy faces, so that must be why they are called flying foxes. I felt that they seemed to be watching us as much as we were watching them, but not in a creepy way.

We went to the shops. I bought a water resistant, disposable camera for the coral reef. Could not help but notice some of the names on the items really do what they say on the tin such as 'open ya bastard' on a bottle opener and 'bugger off' fly spray. The Aussie sense of humour is so straight to the point. We are finding the people are so friendly and upbeat. In fact, later on, a guy at Byron

Bay told us, "Don't take Australians' use of the word bastard the wrong way. It's a kind of term of endearment." He tells us a story where an English guy was called a 'pommie bastard' at work and taking offence, complained to his Supervisor. The Supervisor promptly pulled everyone into his office and said, "Which one of you bastards called this pommie bastard a pommie bastard?"

We are also finding that they love the Scots. We stopped for lunch at a stall which caught our eye because it is a picture of tattie wearing a kilt and it's owned by a Mr McKenzie.

We called in at the Anzac Museum which was very moving. There was a huge number of Australians lost their lives in the World Wars I and II. I knew about Gallipoli and that was about it, so came out a lot better informed.

As twilight came, we saw hundreds, in fact, thousands of wee birds arriving from all over to roost in the trees. Mystery solved! That's what the strange sound was when we arrived yesterday.

All the birds seem noisy in Australia and very, very colourful. The wee birds are beautiful Rainbow Lorikeets and they are so common over here. At first, you think they are so cute as they look like lovebirds, but when they get bored, they end up knocking the stuffing out of each other. They look like sweet, exotic fruit on the trees and then they open their beaks and screech at each other.

We head back to our tent. We just get into our tent when the rain and winds started. I didn't take Priscilla as seriously as I should have. She had warned us earlier that a storm was coming. It was torrential. We were so lucky to be warm and dry in our tent. We only made it by about a minute. Pete keeps getting bitten by insects and my big toe has an infection. Pete has recommended garlic mixed with tea tree oil for my toe which really helps. He says he is going to make his very own midgie/insect repellent.

PETE:

That morning we went into town. As we were walking I realised that my leather bracelet had fallen off. I was gutted. Margaret got it for me at Largs Viking Festival the year before and it had my name engraved in Rune alphabet. As we were well into town we just kept going. No point looking for it. Australia is a big place.

There was an Anzac Museum and Heritage Centre and when we went in they had a film showing about the War. It was quite poignant and eerie as it was done in the style that the soldiers looked like ghosts moving about. So glad I wasn't in the War. We have a lot to be thankful for.

In town, I spied my favourite shop back home: a Charity Shop. They are big into Charity shops in Australia. We had already been in one in Brisbane and I got a pair of Gortex walking boots for $3 which is around £1.80 in GB money. Actually, I am wearing them right now as I am writing this.

As it was so warm we decided to get a hat each. After a good look around a camping/outdoor shop, I got a real bush kangaroo leather hat. As soon as I put it on I jumped over the counter. Margaret couldn't make her mind up between the two she tried on. We left the shop and a good bit down the road I said, "So you decided on that hat then?" to which she replied "eh." I pointed "The one that's hanging round your neck," I said and again she said, "eh," then, "OMG I've walked out with it without paying," and so we made a hasty return to confess and she ended up keeping it.

After that we decided to head back up and as we made our way we saw two grey and pink cockatoos in a nearby park. We went for a closer look and there, hanging on the fence, was my Viking Bracelet. Some kind Aussie had found it and hung it up. I think I am going to like Australia.

Tomorrow we head for Heron Island, famous for the David Attenborough documentary about giant turtles and their young ones running down the beach trying not to be eaten. Our Catamaran awaits…

MAGGIE:

Got the kookaburra alarm call again and I don't mind one bit. I just love hearing it. I know that they are only native to Australia and so I am going to make the most of hearing them. Wish I could take them home with me. I feel Scotland needs a reminder just now to laugh more for we do have a great sense of humour, when we remember to forget about Brexit.

Another lovely coffee and breakfast. Priscilla is getting herself and young daughter, ready for Church and once again I enjoyed our conversation. This time it was about her finding Jesus.

I have never been a Churchgoer in Scotland. I do believe my Church is wherever I am and that the Great Mysterious Universal Spirit is all around us, especially in nature. There are some things which just can't be logically explained and signs are there all the time for each and every one of us. They are all around us. I feel that it has to come from the heart not the head. I am, however, fascinated to hear about different religions and I had said to Pete that I wanted to attend some Churches whilst we were travelling.

Priscilla said she was eighteen when she allowed Jesus in her heart. Before that she said she was on the wrong path. She felt hard hearted and was suffering. She felt she had to soften her heart to allow him in. "When you allowed Jesus in your heart, did it happen straight away?" I asked.

Priscilla replied, "Oh no, it took about three months of constant soul searching and prayer." The Priscilla I know now has an open, cheery face, twinkling eyes and seems a very happy, kind person who is very content with her life. She invited us to go to the Pentecostal Church service with her. Just as she said it, I happened to notice the words GO DEEPER written on a nearby calendar and that is exactly what I decided to do on this trip.

I loved the Church and the service. Everyone was very friendly and casual, mostly wearing jeans and that included the Minister. The really large congregation was made up of all ages and there were lots of families and children. There was a four piece band and three singers and when they started singing it was so joyful and inspirational like a 'Greatest Show' tune vibe. I found myself belting out the words which were up on a big screen. I felt it was coming from my heart. This was very, very uplifting stuff and I wished that more people in Scotland could find this JOY and belief.

The sermon was about how we all have light and dark in us and that Jesus knew this and that is why he did the forty days and nights in the desert. The minister compared having followers on Facebook with the real followers of Jesus. With Facebook not quite knowing who they really are, what they are going to say and what they stand for. Then and now, he said, people knew what Jesus stands for. It was a good way of bringing the relevance of the importance of kind words and actions. Peace and love into our modern lives.

We were invited to stay for a bite to eat and I got speaking to an extremely pleasant Aboriginal woman who I find out, was half Scottish and very proud of her very Scottish name. I spoke to a couple and the lady said she had ancestors from Sorn Castle which is ten miles from where I was brought up. Such a big world and such a small world.

PETE:

The rain and winds seemed to last the whole night. It was wild and I felt lucky we were not out in it. It stopped just in time for the kookaburras to start their morning alarm. I take it they were laughing at us.

Up we got to join Priscilla for coffee. Love it. She said she was going to Church and would we like to join her? At first I was thinking no thanks, then she mentioned there would be a group and gospel music. That was it. Margaret was going to the Church before Priscilla had stopped talking. We had already spoken about trying to go to different Churches wherever we went on this trip.

In the Church, the band came on, bass guitar, lead guitar, drums, keyboard and singers. I ended up clapping along with the rest of the congregation. Margaret was well hooked as it reminded her of 'the Greatest Show'. She sang the theme tune all day, even outdoing the kookaburras.

I had been bitten by an insect the day before. Very itchy and very annoying. It was to be the first of many. I was born with orange hair and was told that somehow this means they are more attracted to me. Hmmmmmm...

MAGGIE:

Priscilla very kindly offers to drop us off at the harbour and we say our goodbyes and take a photo. She is very tall and has to virtually kneel down to appear the same height as us. We both feel like Hobbits. She and her lovely husband have been so kind to us and will have a place in my heart forever. Great start to our adventure.

Now to Heron Island. I am so excited to see the Great Barrier Reef and hopefully turtles, big and small, as well. The Catamaran trip was awful.

I felt so sick I don't remember much about it. I had to go straight to bed on arrival to the Island. No dinner required!

After four hours sleep, I managed to get myself back on my feet and out and about. It was quite late but we end up seeing a turtle in the dark. She was digging her nest, the sand going backwards with slow, laborious thumps. That's twice I've had goosebumps today. Once at Priscilla's Church singing my heart out about Jesus and the other seeing our first turtle up close.

PETE:

So the light had become the word and the word had become flesh and here we were having a real life adventure. Dreams can come true.

After saying goodbye to Priscilla, we head down to the harbour. Can't wait. I have never been on a Catamaran and don't think I have ever seen one. I think I can see the island in the distance, but it turns out that its much further away than we thought. I can tell this is going to be fun. No ordinary ferry for us.

At last it arrives. It's different and bigger than I thought. The people coming off it weren't really saying much. Margaret asked excitedly, "Did you see any baby turtles?" "Yeah" was the reply "but it got eaten by a shark." We looked at each other. That was the first surprise.

Next thing we know the Catamaran was speeding along and it felt like it was taking off. Every time it landed it seemed to go back up higher than the last time. One by one, the passengers started to look green. Seeing them trying to walk was a hoot to watch. I don't think Margaret agreed with me. A guy came up to her looking concerned as she lay flat out on the outside deck at the back. By this time, she had turned fifty shades of white. Since we were Scottish and he was from Liverpool he started talking about Buckfast, as you do. Margaret grabbed the sick bag tighter and groaned. Now I know why nobody was talking as they came off the Catamaran.

MAGGIE:

Set the alarm for 4 am. For the first time since I was a little girl on Christmas Eve, I couldn't wait to go to bed, just so I could get up. The anticipation of catching sight of a turtle in daylight in its natural habitat is palpable. Why? Maybe because it is just so difficult for the babies to get to the water and not be eaten, then once they are in the water the sharks and birds are still picking them off and only a very few survive. Maybe it's because it is a miracle to have such wonderful creatures on this earth. I feel privileged to witness this. We got told the turtles we are seeing are around eighty years old. They come to the same breeding ground but, only do it every few years.

We are out the door for 4.30 am, armed with two tiny torches we got when we arrived. There is a strange ghostly sound all around us. Woooo wooooo. We were told under no circumstances shine torches at the turtles. They are only so we can find our way to our room. It is still dark as we set out to the beach. We are looking for clues.

The sun very kindly obliges us by starting to rise and shed some light. We are looking for tracks and start to walk round the island. Suddenly we see a track. It seems the size of a railway track, so we know that a turtle has made its way up

24

the beach to lay its eggs. We also see a track as if returning back into the sea, so we guess we have missed this one.

We keep walking and within 20/30 minutes we can see the harbour. We have done a circuit. We didn't realise the Island was so small. The sun is creeping up and we keep walking round, sticks in hand. We see a few folk. When we get closer, there is a turtle stuck on the rocks, looking exhausted and forlorn. We wait for the tide to come in. An hour later we get our reward. With tears in our eyes, we are happy to watch her floating away to her home.

PETE:

Heron island is so remote that they don't feel the need to lock doors and don't actually give the option as there are no locks on the doors. We are in the Pacific, miles and miles from land, accessible only by catamaran or helicopter and you can walk round the island in twenty minutes.

We met our guide around 10.45 pm, no torches allowed as it would startle the mother turtle and she would be off. The guides had special lights. As we were walking around the beach there was a loud ghostly sounding noise like wooooooo wooooooo—quite creepy. My immediate thought was voodoo and cannibals. I kept these thoughts to myself in case our guide was one. I wasn't taking any chances.

Finally, we see huge scoops of sand shooting out every few seconds. It actually got me in the face and ears. It was a mother turtle digging her deep nest for the hundreds of eggs, then they eventually become the hundreds of hatchlings. We watched the mother turtle go down the beach and it left a track like a railway track—wide.

We actually became experts in the three days we were there. We were able to tell if they were going uphill or downhill and if the same turtle had caused the tracks. What an experience here in the Pacific watching this real life on the same island that David Attenborough had filmed the turtles and the hatchlings running down the shore. The night over, we went back to our room and opened the Prosecco which we had kept for this moment. We set the alarm for 4 am that morning as our guide said we could hopefully see turtles or hatchlings running down to the sea. We will also, later that day, get to actually walk on the famous Great Barrier Reef when the tide is out. (when the guide wasn't looking we clapped the giant turtle)

Crivens, next thing we know it's 4 in the morning. Up we get as we don't want to miss this. There is that spooky noise again! We see a few railway tracks and being experts by now we decide they are all heading downwards to the sea. We've missed them.

As we walk round the corner we see a turtle stuck in the rocks. It's our first up close and personal meeting in daylight. We keep our distance as she's exhausted, but eventually she gets back to where she belongs. Magical moment.

If you want a tasty, more than you can eat breakfast, I suggest Heron Island. It really does fill you up for the day. It's time for our walk on the Reef.

MAGGIE:

We were exhilarated as we went for breakfast and it was superb. Something for everybody and we sampled it all. Next we are going to hear a talk about Manta Rays, a gentle giant of the ocean. The guide giving the talk obviously loved these creatures and had devoted a lot of time studying them and we came out loving them too.

Next was the Reef Walk. This was one reason we chose Heron Island as I am not a good swimmer and didn't fancy scuba diving. You can actually have guided walks on the Coral Reef here when the tide is out. Our guide from last night is taking it. She is very young looking, very knowledgeable and the guided walk is incredible.

The Great Barrier Reef really is a wonder of the world. The guide shows us three types of sea cucumbers which she picks up and lets us touch, two stingrays—"if you don't annoy, them they won't annoy you"—we were informed—starfish, turtle moss which is poisonous to every creature except a tiny green crab, clams, lots of lemon sharks (harmless) reef sharks (not as harmless but docile). The star of the show was really the different corals. "Watch your leg doesn't touch the coral it is sharp and can infect." Too late Pete had a scratch on his leg. Not to worry Sam has the antidote—iodine—and he is fine. Sam also explains that much of the coral is bleaching its colour because the sea is heating up due to global warming but studies are taking place on Heron Island to see if this can be reversed.

I have noticed that Australia is not bogged down by the health and safety that ensnares the UK and the USA, which, let's be honest, is not really to protect us, but just in case someone sues. It is a killjoy. Long may this continue in Australia because we chose to go this guided walk and that should be enough. What we have just witnessed was so fabulous, we feel so lucky and filled with awe. As our guide said if you don't annoy the creatures then they won't annoy you.

PETE:

We get shown a film about Manta Rays and sharks then we go walking in the real coral. It is very colourful. We get shown how to identify sea cucumber, starfish. We got told earlier DO NOT REPEAT DO NOT pick this up. It looks like a shell and is in fact very poisonous and can kill nearly instantly. None of us pick up anything, we leave that to the guide.

Suddenly I see two stingrays. I freeze, then I freeze even more when I realise I have cut my leg on the coral. F*** s*** I say to myself have I attracted the rays with my blood? Well apparently not. I am relieved to find out they are docile. Our helpful guide takes me out of the water and I get iodine rubbed on my leg since corals have bacteria.

Next day I decide to go snorkelling. There is a crowd of folk doing it so why not me?. So off I go for the safety instructions. Turns out these are my kind of instructions:

Instructor: Can you swim?

Me: Yes.

Instructor: Good. If you get into trouble, wave your hands in the air and shout for help. Have a great day.

Once I got my snorkel, I decided to go further away from the crowd. Kinda out of sight. I had never snorkelled before and as soon as I get in the water I swallow a big mouthful and when I try to stand up my feet don't touch the ground. I try not panic and being a Taurus, my next step forward my feet touch the ground. I look up and I am only a couple of feet from the edge of the water. I take a deep breath and decide snorkelling is not really for me. I thought you just floated on the water as everyone else seems to. I decide to go back in, as Margaret is watching and I don't want to look like a wimp, so I gave her a wave as if I was just coming up for air.

MAGGIE:

As Pete came out the water in his flippers he looked as if he was auditioning for a part in Mamma Mia where they guys do a snorkel and flipper dance to Lay all your love on me. LOL. I wish I had my camera. Speaking of cameras we couldn't get the underwater camera to work yesterday. Think the guy in the shop saw me coming.

It is bird heaven here. When we arrived, I heard that there are two rare sea eagles on Heron Island and some sacred kingfishers and shearwaters. It is my wish, as a lifelong bird lover, to spot these rare birds. A couple sitting nearby point and there is a sacred kingfisher sitting on a rock by the jetty, colours glittering like a jewel. A wish come true.

We get up again this time at 4.30 to be out for 5 am, when the sun starts to come up. Still hearing the moaning and groaning, wooo wooo sound. Pete is going to do his Shaggy from Scooby Doo impersonation and solve the mystery.

We feel like pros now and start circumnavigating the island, looking for the tracks. There are lots of tracks, many of them showing that the turtles had been up, did their stuff and back down again. We then saw a long track going back to sea and there is the turtle twenty yards to go. It is such an effort for them out of the water. The sun is rising, we are on a paradise island and get to see this miracle. We wish our turtle friend well and bon voyage.

The sea eagles didn't show up until 6 am when they put on a rare display for us as we walked round the island for the umpteenth time. I was about to see wee hatchlings running down the beach. The way it was timed we only got to see a few as most of the eggs had hatched a couple of days earlier and these were the stragglers. The guides are unearthing and counting the unhatched eggs for their field study.

As one tiny wee thing struggled to get over the grass, I realise it really is a monumental journey for them. The seagulls are gathering like vultures and we are not allowed to touch the baby. We do our best to shoo away the gulls and the little one gets to the water and goes out for about fifty yards then a gull swoops down to go away with it. That is why they need to be in their thousands when they hatch. Mother Nature can seem cruel but we have to respect her, as she knows about balance much better than we do.

I have had all my wishes granted on this wonderful trip and now it is time to go on the glass bottomed boat to see the Coral Reef in all its glory under the sea. A kaleidoscope of colours. Beautiful, precious, living corals; such variety of fish: clown, zebra, angel and parrot to name some and we can recognise sea cucumbers and clams and stingrays because of our Reef Walk yesterday. Then I see it and it just melts my heart. We see a large sea turtle swimming, with such grace and elegance, in its natural environment. No longer clumsy and slow; such a majestic creature, it looks so happy. It just does not get any better than that.

PETE:

Well, that's enough snorkelling for me and as I head out the water I nearly stand on a shovel head shark. I repeat the mantra: if I don't bother them they won't bother me.

We are going for an underwater boat trip in an hour or so. We decide to go for a walk round the island again. You don't seem to hear the spooky wooooo wooooo in daylight. As we are walking there is a couple on the beach reading. We get talking and the guy asks us if we are Scottish? He hands over the book he is reading. It's Billy Connolly's autobiography. It feels great to be Scottish. I remark that I once spoke with Billy for nearly an hour, just the two of us… As we left, the guy called me wee Billy and invited us to pop in to see them on the way back to Brisbane.

Time for the boat trip and boy does it live up to our expectations. As we arrive back in the harbour a huge shout goes up. THERE IS A HAMMERHEAD SHARK, IN THE HARBOUR, ATTACKING A STINGRAY. Right in the middle of all the snorkellers. The coastguard shouts DON'T GET OUT OF THE WARRER, THIS IS RARE, TAKE A PIC! Two minutes later everyone is swimming around again. I just love the Australians. They just get on with it. They live with these amazing, sometimes dangerous, creatures every day. If that was back in the UK, the beach would be shut for weeks, even though the shark only stayed for a couple of minutes. Hammerheads are pretty docile with humans but I was kinda glad I had handed in my snorkelling gear.

We got up at 4 am again since it was our last morning. When we opened the door, the spooky, ghostly noise seemed very loud. I grabbed a torch and looked around, determined to get to the bottom of the mystery. Woooooooo wooooooooo. I shone my torch in the direction and there they were: lots of bird couples. They were the size of gulls, with long beaks and they were sliding their beaks up and down each other's beaks. It sounded like a lament. They do this every night to each other until daylight. They are called shearwaters and live in burrows in the ground, They don't see each other in the daylight and it's as if they are lovesick. How cute is that. I feel like crying it seems so sad and yet it is very touching. At least, it means there are no cannibals and I can talk to my guides without fear.

One of them looks about eighteen, but is in fact twenty five. A qualified Marine Biologist, she wishes she looked older so people would take her seriously and ask her serious questions. I am sure she will be fine. She knows her stuff. You know just by listening to her, I say.

Well, we leave the lovebirds woooing to each other and head down to where, being 'experts', we think we will see the baby turtles and see them we do. They are scampering through the reeds, getting caught, turning over on their backs. They really do have a struggle and we are told we can't intervene, but Margaret is already waving her arms at the seagulls and speaking in what can only be described as deep Scottish 'GETTAEFFYABAMPOTS'.

Another couple of circuits round the Island and we saw another couple of mother turtles returning to sea. We feel all smiley, smiley and happy when we look up and see not one, but two, Sea Eagles. This place is paradise. I will never forget it as long as I live.

Next morning we both say goodbye to the turtles, the hatchlings, the sharks, the woo woos, the sea eagles, the breakfasts, the palm trees and the coral. I feel like I'm mates with David Attenborough. WELL HERON ISLAND: NOT MANY PEOPLE GET TO DO THIS, SO THANKS and as we waved goodbye we didn't realise just how many more incredible experiences were to come.

Rum, Ginger Joe and Gin Gin

PETE:

As we got off the Catamaran and back on dry land, we did the same as everyone else we saw coming off, when we were waiting to go. We gave a look to those waiting that silently said—that was brilliant but you are in for a bumpy ride…

So we are ready to hire a car in Gladstone. Burleigh Heads here we come. There are two reasons we are going to Burleigh Heads, one of them is because one of us is called Burleigh: ME.

I am allowed to drive the hire car. I passed my driving test last year at age sixty three (so go for it and don't wait like I did). After a stop at Mirium Vale. We have a walk and an ale and then we get up early and on the road again.

After driving for around an hour without seeing much, something catches our eye. It's a big building with huge writing on the walls saying FREE TEA

AND COFFEE—HAPPY NEW YEAR. In fact it used to be an old Railway Station from way back. In we go.

"I see your sign on the wall. Can we get a coffee and tea please," I say, "how much is it?'"

She looked alarmed and said, "It's free, tells you on the sign on the road in mate, only we're closed for half an hour yet so you'll have to go help yourself in the kitchen, put the kettle on, milk in the fridge, tea and coffee on the shelf."

Her name was Sheila and we fell in love with her frankness and honesty. She also had a free juke box with loads of Led Zeppelin songs on it. How cools that. Even cooler, there was a rug over the pool table. I asked if it was a kangaroo.

"Nope," she replied, "it's a wildebeest. God knows how it got there. Some drunk left it. Been there ever since."

Sheila was interested in us and gave us information on a couple of free camping Apps which came in very useful. Sometimes she says, all you need to do is buy a couple of beers at a pub and you can park in the car park grounds. Good advice. Thanks Sheila. Before leaving, I ordered a fish cake burger and to my good surprise it's not pasty or soggy its real fish on a roll. The fish is called Barramundi, Bye Sheila. Love you :)

MAGGIE:

We decide to hire a car and drive back down to Brisbane airport, then catch a train to Burleigh Heads. This will give us the freedom to explore the Sunshine coast, as we already did the train journey on the way up. We hire an Estate car which will allow us to sleep in the back of the car. It's only for four nights.

Heading down the Highway we notice straight away that the roads are great. They are wide and well maintained. Sometimes the Highway runs right through the middle of a town and we decide to stop at such a place—Mirium Vale—for the night.

We hadn't been to an Australian pub yet and decide to give one a try.

"Do you want a Stubbie, a Schooner, a Pot or a Pint?" the barmaid asked.

I ask her to please explain. "A stubbie is a bottle of beer, a schooner is less than a pint but more than half a pint; a pot is half a pint and a pint is a pint." I decide to go for the schooner of ale. Pete does the same.

The pub is not that busy, but there are maybe ten folk counting us. About fifteen minutes after coming in, around 8.30 ish, the barmaid calls TIME TO GO

HOME. Seems pubs have the discretion to close whatever time they decide if it is not busy enough.

She does us a favour because we decide to get up early around 5 am. We have noticed that because of the heat folk get up really early, go jogging or whatever while its cooler, sit out for breakfast etc., and then are in bed by around 8 pm.

Apart from one tropical storm at Gladstone there has been no rain at all. My hat is proving to be a wise investment. It is around thirty degrees every day. Luckily, there has been hazy cloud and with our hats on we have been able to function really well.

Got up early and headed for a walk around Mirium Vale. Saw more Fruit Bats, kookaburras, butcher birds. Then we are driving on, the hire car is automatic everything and very, very enjoyable to drive.

As we are driving Pete notices a diner and we pull in. Sheila is the kind of Australian I expected to meet all the time. Frank and funny and not at all politically correct.

We ended up staying for three hours and having a chinwag with Sheila, a shower for 3 dollars and as much tea and coffee as we want.

When we said we were heading down the coast to Sydney, she said it was full of Towelies.

"Towelies?" we asked, perplexed.

"Yeah they wear towels on their heads." She just said it so dead pan with no malice. We came to realise that Australians love to give anything and everyone a nickname. Sheila made our day with her kindness, hospitality and entertainment.

We arrive at 1770. Not the time, but the actual name of the town. 1770 is smaller than we thought. It's where Cook landed on his second visit to Australia in the year 1770 and is famous for its sunsets. Unfortunately it was too cloudy this evening, but there are lots more sunsets to look forward to. We are sleeping for 8 pm with the sound of the lovesick wooooo wooooo birds in our dreams.

PETE:

The estate car is an automatic. I love this: the lights come on themselves when it is dark; it changes gear all by itself, so no farting about at roundabouts or traffic lights. The roads are great. They are wide, long and straight.

Some of the trucks are real fancy, shiny colours, so I take a picture to show my grandson, Taylor. He would love them. They even honk and toot as they pass. I think this is great until somebody points out that I was probably hogging the road and lucky I was not knocked off the edge.

As we go down a country road something runs across the road ahead of us. It looks black and stocky with short legs and it can move pretty quickly. I think I have just seen a wild boar.

We arrive at 1770. It's the only place in the world where its name is a number. It's named to commemorate the date Captain Cook landed. Captain Cook went on to land on Hawaii where they liked him that much they ate him.

We went into a small rainforest here and saw a couple of Bustards, yes Bustards not the other word. They are big birds like turkeys. When Cook first saw them, he killed and ate them before he went on to Hawaii, Bit of heavy karma going on there. Maybe his name should have warned him.

We come to a place called Bundaberg and have my first encounter with the Australian police. Luckily I had parked the car and was walking through a red light area. No not what you think, but traffic lights. I got a look and nothing more, then we see a sign for a Distillery: Bundaberg Rum.

In we go and get shown round. We are offered to make our own rum flavour and if it's a winner they bottle it and sell it. We decide to use our two free drinks tickets that you pay for in the entry price. I have rum and vanilla and don't know what Margaret has. She looks as if she is not keen on it, but necks it anyway and doesn't even bat an eyelid. Mine made me lift my heel and say, "Shiver me timbers."

As we leave to wander the streets we come across a Brewery with all sorts of real ale. We sample some like Big Yak, Rusty Roo, Hefty Heffer and my favourite Drunken Fish which tastes way better than it sounds. Since it is always dry and warm here we have been drinking a lot of Bundaberg ginger beer and we were fortunate to come across the Ginger Beer factory which we enjoyed better than the rum, It's a great family day out.

If all this talk of rums, ales and ginger beer is making you thirsty, the next town we come to is called Gin Gin. Only there is no gin distillery. Every town has a bottle store which sells purely liquor and booze. We entered the bottle shop and 'HELP MA BOAB' there in front of us was ginger beer, alcoholic ginger beer, which you already know by now: we love ginger and we love beer.

This one was called Ginger Joe because the man who invented it loved ginger, had ginger hair, was called Joe and according to legend when he sampled his first batch it singed the twirly bits at the end of his enormous moustache. "That'll do for me mate," I say and he was to become a good friend of mine in the heat of Australia!

I was to come across another red haired legend. I first noticed him earlier on our trip, on a guy's arm in the form of a tattoo. It said The Wild Scotsman and had a face with loads of red hair on his head and chin. In Gin Gin, I saw a poster and I recognised the face. Turns out there is a museum in his honour and he was quite a character. His name was James Macpherson and he was a Bush Ranger (not a Lone Ranger) he was an outlaw and skilled in bush craft.

Despite its name, I think Gin Gin is a Christian town as when we got up the next morning, Sunday, I noticed there was a gospel service offering lunch and tea. I tell Margaret, as she is always looking for gospel singing to let rip and an excuse to dance about. She makes my mind up for me and tells me we are going.

I didn't enjoy it as much as Priscilla's church but I did enjoy lunch and got some help on my quest to make a midgie cream as I discover yet another ingredient that the Aborigines use. That will do for me. My cream is getting nearer. Still being bitten every day.

MAGGIE:

Heading down the road towards Bundaberg we notice it is a very agricultural area with lots of plantations. Fields of sugar cane, fields of macadamia nuts and fields of ginger. We have been drinking lots of non-alcoholic Bundaberg ginger beer as it is so thirst quenching in this heat and we both love the flavour of ginger. In fact, Pete is the only other living person I know who loves ginger as much as me.

We end up in Ginger Heaven as we come across a Bundaberg ginger factory, with all things ginger. We had so much fun there and got to see where our favourite non-alcoholic ginger beer is made.

Before we found this, we had stumbled upon Bundaberg distillery which we thought was the ginger beer, but in fact it was Bundaberg rum. That explained all the sugar cane in the fields. They turn it into molasses, the main ingredient of rum. Rum is not my favourite tipple but we find out that it really is big in Australia. I think it's in their DNA or psyche. When the convicts first landed with James Cook, there was hardly any food but there were lots of barrels of rum and I don't think they've stopped drinking it since then.

Met an interesting couple from Adelaide, up to see the turtles. Turns out she is from El Salvador and they met through work. They are very pleasant and we talk about USA and their South American propaganda and also why we agree that one should not put ice in rum or whisky if one is a connoisseur and wants the real flavour.

We went walking in the city. It has one hundred thousand residents and it feels like ten thousand. Brisbane is the same. There seems to be lots of space between the houses and the roads are wide. We end up watching Boogie Boarders at the beach and then Pete suggests we head for Gin Gin. So good they named it twice.

Gin Gin is across country and the Bruce Highway runs right through it so it is easy to find. The roads are such a delight. No pot holes. I have also noticed that Australian public toilets are great. They are often open twenty four hours and quite often it is volunteers who clean them. They are mostly clean, fresh and welcoming. Every town has one. Australia is very camping friendly.

Gin Gin has good vibes. When I wake up, I realise it is Sunday and just at that Pete comes back from a walk and tells me there is a gospel church. As we go in they are starting to sing O Happy Days and I immediately join in. I enjoyed the sermon which is all about having a healthy tongue. Watching what words we say, being kind, be careful what we listen to and have healthy ears; being true to ourselves and have a healthy heart. He could quote bits of the bible to back it up. I thought of the well-known saying 'if you can't say something nice don't say nothing at all'.

Such lovely people, wonderful hospitality, they invite us for lunch and it's all homemade. The keyboard player, is moving soon and invites us to contact her at her new home near Brisbane, giving us her mobile number, as does the Minister who also gives me a large smiley face sticker. This ties in with his sermon and with him too, as he reminds me of Neil Sedaka who also has a smiley face. I got a lot from his sermon.

Dodgy O'Doyle, the Dead Cat and the Tiki Tiki Woman

PETE:

Well, we made it. It was five days and four nights in the car travelling down the Bruce highway. Bruce was the name of Margaret's husband who had passed away two years ago after a long illness. Although they were separated at the time, they were still good friends. As my partner Isobel also passed away at a young age, we both knew how it felt and realised how lucky were to meet each other and have a second go at life. We will never forget them and both feel they have both been on this journey with us.

Here we are at Burleigh Heads and the second reason we are here will be revealed shortly. It is a fantastic place and it feels quite odd being in a place called Burleigh as my name is in lights everywhere. Burleigh Hotel, Burleigh Cafe, Burleigh shops, Burleigh clothes.

I was talking to a cafe owner and seemingly Burleigh is quite a rare surname so, when I told her my name she beamed, bowed to me and said something like 'honoured mate'. I felt like the King so in return I tipped my real Australian bush kangaroo hat and said 'G'day' and 'Rise'.

Later that night I felt 'uneasy'. I stopped a guy and girl and asked, "Where is the best place for a good coffee?" The girl was helpful and quite chatty then I mentioned we were on holiday and my name was actually Burleigh. At that the guy got out of his car. He had a stick and was limping. He says that his name is O'Doyle and that his friends will never believe him that he met an actual guy called Burleigh and could he photograph my driving licence as proof and would we like to go to a club with them. I politely say, "NAW" grateful never to see him again anywhere.

Next day we decide to go to see an area called Koala Park to see if we could see a koala in the wild.

It was on our way there that one of the weirdest things on our travels happened. We came to a path that led to a rainforest. I have always wanted to go to a rainforest since I was a wee boy watching Tarzan films, so we ventured in.

Then: bugger me, mate, there in front of us, out of nowhere, appears this wee woman. She says she was bitten here by a poisonous brown snake four years ago and to be careful. Then she said she had come from Bethlehem and Judea and that during her research she found out she was in fact from the line of King David of Judea. Then she started to speak in 'tongues' something like 'Hikatta, Heekatta, Katta, Kattah, Heekatto Khom Keetakhom', then she put her hand on my head and said I was going into the heart of the enemy but do not be afraid as Jesus has been there and I was OK, nothing to fear, just have faith. Fair enough: I have plenty of faith.

THEN! She put her hand on Margaret's head: same thing 'Hikatta, Heekatta, Katta Kattah…thingmabob'. Then she told her she was a woman of valour and she would lead people, she then blessed us again in English, spoke to Margaret then 'tongues' again. Then as we looked at each other and back at her she had disappeared. Never saw her again and never got to see a koala that day.

MAGGIE:

We are heading to Burleigh Heads via Herley Bay on the Bruce Highway. We take a moment to think about Bruce and Isobel our previous partners, who have both passed away and to thank our lucky stars we are still here. As we pass through Childers I spy an ice cream shop which is in an old cinema. We get the last two macadamia and ginger ice creams. A pleasant young girl who served us invited us to go and look at the old cinema. Very quaint with a piano. All that was missing was an usherette.

The run down the Sunshine Coast was lovely but now it's time to hand in the car and get the train to the Gold Coast and Burleigh Heads. We have booked another Airbnb for a couple of nights.

We immediately love the place. We see the name Burleigh everywhere. It is great to see Pete's name in lights in the evening.

We go for a walk and end up at the Aboriginal Centre and the sacred hill rainforest called Jellurgal. As we were walking along the footpath amongst the trees a wee woman stepped forward. She had a twinkly face and eyes and seemed very friendly. She explained that she was originally from Australia but had been in New Zealand for years and now the Cook Islands which is very remote. She was over a visit and was reminded that four years ago at this very spot she was bitten by a venomous brown snake. She had been warned not to go down this path but she had not listened and was bitten. Luckily she survived just like Paul in the bible. She put her hands on Pete's head and said, "You are going into the heart of the enemy" Kia Tiki Tiki something or other she then placed her hands on my head and said I was a warrior like Deborah in the bible and that I had to face all of my fears one by one Kia tiki tiki something or other. She said I was to let Jesus into my heart and that I had to use my healing hands. I felt really safe with this woman and so did Pete. She then said something about she was from the line of King David and mentioned the star of David. I thought now you have gone too far lol. Pete and myself kept looking at each other and smiling not believing that this was happening. We looked at each other again and then when we looked back she was gone. Well that was unexpected.

PETE:

It's 13th February and today I got my hair cut and my beard trimmed at a hairdressers across from the Aboriginal Culture centre. The woman's name was Dianella. She was very nice and when she was washing my hair she kept going away and coming back again. Turns out her cat had died the night before and she had to go and cry: fair enough, but does my hair resemble a cat? It was a very nice feeling though, when she was doing my hair, it felt as if she was stroking her cat, so I hope that I helped.

I notice that the local Aboriginal name for Burleigh Heads is Jellurgal which I guess means my new name is Peter Jellurgal. The reason I got my haircut and beard trimmed was also the second reason we are in Burleigh Heads. Tomorrow is February 14th. Any ideas yet? Valentine's Day. Ring any bells? Yes, well guessed. We are getting married. Here in Burleigh Heads, Australia. Two become one on St Valentine's day.

We know; it's cheesy enough getting married on Valentine's Day, even more cheesier it's in Burleigh Heads and EVEN more cheesier than cheesier we are getting married on a beach (Burleigh Heads) in our bare feet, me with a Beatles T Shirt and Shorts 'All you need is love' and a flowery, flowing dress for the new Mrs Burleigh.

Well this all added to the adventure. It wasn't deliberately planned for it all to happen this way. It just did. We knew we were going to Burleigh Heads, as the trip got nearer we realised we would be there on Valentine's Day and the rest, as they say, is history. We realised we love each other and that doing this was inevitable.

MAGGIE:

We didn't think of getting married when we booked the trip but then when we saw Burleigh Heads on the map, We thought that's it—the perfect place for me to take the name Burleigh. We decided to keep it a secret from family and friends. I found a fabulous Wedding Celebrant online.

We are going to meet her at her house the next day to finalise our documents etc., and talk about the ceremony which will take place on 14th February 2019 at 3.24 pm.

We immediately like our lovely Celebrant, a warm hearted Kiwi who has lived in Australia for many years. She is going to provide witnesses. She tells us many couples get married at her flat, on the balcony. It has such a lovely view.

We decide that we would like to get married at the Aboriginal Hill. Our lovely Celebrant advises us to walk around later and find a spot. She recommends a spot on Little Burleigh Hill. We ask what should Pete wear as I have brought a simple turquoise dress with me which is appropriate for the beach wedding. She advises wear what you are comfortable in. Pete decides on his Sergeant Pepper T shirt. All you need is love. He is so happy about this and he will go and buy a new pair of shorts tomorrow. We say our goodbyes until the 14th and will text to let her know our chosen location.

When we take a walk to Little Burleigh Heads, it feels right. Our lovely Celebrant had mentioned that many couples have put a padlock on the railings symbolising their love. Pete happened to have a padlock with keys. (from his rucksack I think) We decided this was the ideal location to make a vow to give each other the freedom of true love. I looked at the open padlock in my hand. The key to true love and my heart, is freedom. An open padlock symbolises this to me so, we decide not to bother with locking a padlock. We find out later we couldn't get married at Jellurgal, the big hill, anyway as it is a sacred Aborigine site, which we totally respect.

We went shopping for a pair of shorts for Pete and find the perfect ones to go with his Beatles T shirt. We decide to get married in our bare feet, but buy a pair of slider sandals each to get to the beach. I am going to get a pedicure and have my toenails professionally painted for the first time ever. We just need to find a hairdresser.

We walk over to where we will be spending our wedding night. We are allowed to take our things in early on the 14th and get ready there. It's at a place called Koala Park and it looks amazing. There is a row of shops and I spy a hairdresser shop. I walk in and our song is playing on the radio. It's the first song we ever danced to and it's called A Good Heart by Feargal Sharkey. I take this as a sign. Her name is Dianella and she is a lovely, gentle soul. Pete puts an appointment on today and I for tomorrow.

All You Need Is Love (Rah Tah Tah Tah Tah)

PETE:

It was a beautiful wedding. All we could have asked for. We got married on Little Burleigh Hill next to the beach. We had special rings made. Silver ones with runes engraved on them each symbolising 'Dreams of Everlasting Love' and they were forged in Orkney in the land of the Vikings by Sheila Fleet, a marvellous jeweller. They are very, very cool to look at, so that is another cheesy thing we did, although, without a doubt, the best thing was the actual vow itself. Not for us the mainstream 'I Do'; oh no, we went even cheesier and said these words to each other: I AM READY... I AM WILLING... I BELIEVE. Then exchanged rings; BEAUTIFUL!

I had my Sergeant Peppers t-shirt on and Margaret looked stunning in a lovely sort of aqua blue flowing, flowery dress. I always like looking at her anyway, but today somehow she looked like a heavenly angel brought down to Earth just for me. We walked down onto the beach and with our toes in the sand, hair blowing, we held hands and walked along the golden, sandy, beach with the odd wave reaching our toes. Now that's cheesy.

We didn't have to look far for our wedding meal. The first place we came to had a big sign up saying MENU. I said to Margaret that's us (Me n U) then below it said RUMP AND A PINT—FIFTEEN BUCKS. Well, you don't have to say that twice, mate.

Later as we walked along the beach towards the rainforest path, we came to a spot of green grass, a sort of park if you like, and a notice that said NO ALCOHOL ALLOWED, but to our very pleasant surprise there were dozens of couples sitting quietly in twosomes with candles lit and glasses of wine, celebrating St Valentine's Day. It was a lovely sight and just being married, we sat down in our twosome and joined in with the rest.

MAGGIE:

The big day has arrived. It is another lovely day in Australia. At breakfast, a waitress commented that I look like Doris Day. I take that as a massive compliment. I feel so happy. No nerves. Just looking forward to this day so much. Dianella makes a great job of my hair and my toenails look amazing too. I put my own make up on. We both had a great time on our wedding day. It was everything we ever wanted and more.

Our lovely Celebrant had, as we asked, put the anonymously written True Love poem, which we came across at a visit to Gretna Green, into the beautiful service. We both had big smiles all the way through the ceremony as we said our vows: I am Ready, I am Willing, I Believe.

The two witnesses are great characters. Cathleen, an expat of thirty years, took some wonderful photos for us. Bernard is a real medallion man. An 82-year-old former opal miner, he used to live underground because of the sweltering heat where he mined. We all went for a meal afterwards. Our lovely Celebrant doesn't usually, but Bernard was hungry and it was lovely to share our day with them. We went to the nearest restaurant who were having a rump and a pint for 15 dollars. Nuff said lol.

It feels great to be Mrs Burleigh at Burleigh Heads and we walk back along the beautiful beach hand in hand. A young girl and myself smiled at each other. We could feel the love in the air. Happy Valentine's Day she said. Same to you I said and told her we just got married. She was so delighted for us and so was her boyfriend. We walked up to big Burleigh Hill or Jellurgal to give its proper name and lots and lots of couples were having picnics to celebrate Valentine's Day. Love Burleigh.

PETE:
We didn't want such an out of the blue thing to end, so before it did we headed to our honeymoon suite room and boy did they pull out all the stops. It was pure luxury, giant four poster bed with white silk and lace curtains, lovely lighting, chocolates, brandy and our own private pool and jacuzzi. We woke up and we were Mr and Mrs Burleigh. Margaret wanted to stay the full day lol. We did, apart from a walk in the woods looking for Koalas. So thanks, Margaret, you are my dream of everlasting love and I love you. Pete xxxx

MAGGIE:
Woke up early in our beautiful room at Koala Park Estate and we decided to go looking for a koala. I buy an ice lolly and as I am giving Pete a bite, a young girl walks past and says, "You two are so cute!"

Koalas can be very shy, but after taking advice from a couple walking in the woods, we took our time and we spotted one. A lot of Australians have never had that privilege. So exciting to see animals in their natural habitat. It saw us too in the distance as it's ears perked up and it stretched its neck to look at us.

A nice start to our honeymoon. Then we went back to the apartment and lay by the pool all day and soaked up our surroundings as tomorrow we are going for our camper van.

Before we head back up to Brisbane Airport for our camper van, we popped in to see Dianella our Hairdresser to give her a card and say thanks. We showed her photos and told her about our big day. I think we 'nudged her stick' (ask Pete about that). We have, in a short space of time, got to know her quite well, sharing confidences as hairdresser and client often do. She is a gentle soul who is learning to stand up for herself. I said, "You are a woman of valour." (Kai tiki) as I left.

Forty Days and Forty Nights

We love our campervan. He is orange and white, with the word Hippie surrounded by flowers and his name is George. We realise that we have the van for forty days which we did not do intentionally. When we study the map, we decide to head down the coast to Sydney and back up the Hinterlands to end up back at Burleigh Heads six weeks later. Then we go to Brisbane for two nights before catching our flight to Christchurch. We feel we will be able to see the real Australia by taking our time rather than trying to tick too many boxes.

Firstly we are heading down the Scenic Rim road to a town called Boonah. We find a Showground which is a good way to camp in Australia. It's big and it's busy. There is a caravan club visiting and lots of tourers, but also a lot of residential caravans too.

We asked a guy for advice about our leisure battery and crikey ended up we got his life story. Seems that he had an acrimonious divorce and it so happens his wife and her whole family are either lawyers or judges. To cut a long story short he deliberately drove his truck into the front doors of the house that he built and it took four hours to dig him out. They couldn't charge him because it was his house, but now he is in hiding. Bit of the Dodgy O'Doyle about him if you ask me. We were none the wiser about the battery and I was glad when a guy came up to see what was going on and join the chat. I was mighty relieved to eventually get into our campervan and lock the door for the night.

PETE:

We saw a koala in the wild yesterday at Burleigh Heads and we decide that is our first wedding present to us from Australia and boy would we get plenty more. It is time to travel. Time to have our honeymoon, in a campervan, touring Australia for 40 days and 40 nights (it was a surprise when we realised this was the time we would be in George) He is a colourful van with HIPPIE written on it in a flower power style. He feels good, looks small, but then again so do we.

There are loads of vans on the road so we are not the only ones living the dream. First stop is a place called Boonah. I decide to ask a guy if there are any wombats around but to begin with I ask if he knows if we are charging the leisure battery up correctly.

Well! I didn't get a chance to find out about wombats. He was quite helpful, but Aussies like to talk and the next thing we know: his wife has left him; all her family are involved in law; so he explained he was left with nothing, it's all gone, even his house phone was blocked; he was that angry he drove into his house. If he were us, we'd be better going to fkn Victoria, there is nothing here mate. My mum has been waiting five years for rain.

I was about to say come to Scotland mate, when without stopping for breath he says the government cut down all the trees mate so there is nothing to attract the clouds, so the clouds move on to where the trees are. The fkn idiots that's the fkn government for you.

I must admit it came as a bit of a surprise that here we are at the start of the Scenic Route and there are no trees. We decide to chance it, quite sure we will see something as it is called the Scenic Route, but not before a guy comes up and says G'day there's more than two folk here so I figure there must be something going on.

Next day we met another guy and tell him we are starting the Scenic Route. I have lived in Australia for years. Here is some good advice: always camp at Showgrounds if you can, you get power and they are cheaper than campsites. Plus you get to see guys with horse drawn buggies and if they hit a stone they go five feet in the air. He then came back with an expensive looking book on free campsites, pubs, showgrounds, maps etc. He was a real cool dude, so off we go on the Scenic Route waving to our generous friend as we left. Bye Boonah—toot toot, honk honk.

MAGGIE:

Since it is Sunday and we are up early, living like the locals, I decide to go to a nearby Church. This church was not busy and I was one of the youngest there. Lots of saying prayers en masse but in a quite dreary way. The sermon was about generous hospitality (Luke) and forgiving generously.

The congregation seemed friendly, all shaking each other's hands saying peace be with me including me. However the irony is that the only person who ignores me is The Minister. She deliberately walks straight past me. I think that could be because to my great alarm earlier in the service my phone alarm had gone off. I realised I forgot to cancel it when I woke early. I think she was finding it difficult to forgive me or to welcome me generously. Time to leave.

We head on our way and end up at Moogerah Lake which we find out is a dam.

It was a beautiful day, very hot, so we park in the shade. Two busloads of Chinese people arrive and it just happens to be the Chinese New Year. It seems very busy. I shook hands with a couple and wished them a Happy New Year.

We were overjoyed to see an osprey fishing on the lake. Pete decided to go the three and half mile round trip up to the top of Mount Edward. I decide to stay at the bridge to take in the idyllic view.

I thought I saw a duck billed platypus. I really would love to see one. I did a talk at school at the age of eight and I chose the duck billed platypus as my subject. I have been fascinated with them ever since. I ask a young couple did I just see a platypus? The young fellow said I have been here for twenty six years and have never seen a platypus. Turned out to be small, fresh water turtles.

PETE:

Driving along enjoying the great weather and stunning views we decide to stop at Lake Moogerah for a better view. I decide to reverse only to hit a big 4 x 4. I was a bit shocked. I got out just as a young Aborigine girl got out and just stood looking at me.

Then a big Aborigine guy with a big Aborigine friend got out of the other big 4 × 4. They walked up, looked, then the big one said to the other big one, "Well, what do you think?"

"Hmmm, looks like a scratch on me tow bar that's gorra be about two thousand bucks worth of damage." I just stood there and then they both burst out laughing. Big infectious laughs and asked if I was OK. I like the Aborigines already, little did I know that one day I was going to end up attempting to play the didgeridoo with real painted guys.

We walked up Mount Edward, well I did. Margaret decided not me this time. She decided to enjoy the marvellous view. She was on a bridge and down below were small, fresh water turtles. I only saw one, because I was up a mountain and they had mostly gone by the time I got back. As we were looking at the water a large bird with huge wings swooped past. I knew from school it wasn't a Pterodactyl. It was in fact an Osprey. This had turned out a lovely and rewarding day.

We decided to use our new campsite book which our friendly Aussie had given us. We do a lot of hmmmm I think it's this road only to see a sign that says turn off your sat nav as it doesn't work here. We look at the book again and it says camping ground and pool 5 bucks. Well, that is cheap says us and off we go quite happy with ourselves for being good map readers.

Sometime later as we are driving along I notice a field full of old, rusty tractors and cars and a big oil slick kinda hole next to it. I said to Margaret. "That looks like an old farm let's ask that guy who is waving at us which way is the campsite." Margaret says, "Ok I'll drive up to the bend and turn." When we get to it, a notice says Stop Dead End and an arrow pointing back down the way. We then see a sign saying campsite. It is the field with the rusty vehicles in it and the oil slick is the pool.

We assume that the vehicles once belonged to innocent travellers like us who went for a swim and couldn't get back out. We decide to leave just as the guy was getting closer, so we put our foot down and somehow ended back at Boonah.

So much for the tips in the camping book. After we had calmed down, Margaret decided to have an Indian Curry since we were in Boonah it had put her in the notion. I went looking for a Chinese Restaurant and found 10 dollars, so I used that.

MAGGIE:

After looking at our new book, which I think is quite old, much discussion we decide to try the Gorge Campsite with pool. The Sat Nav didn't seem to know where it was and took us 20 kms wild goose chase, when it turned out it was only

5 kms away from where we started. I think it was trying to warn us. I was adamant we were going because of the time it had taken trying to find it.

Once we found the 'campsite' up a dirt track, it looked horrendous, like something out of a horror movie, hence we made a sharp exit and hence we ended up back in the safety of Boonah for another night. I had an Indian meal as the name of the town had given me a notion for one. Everything was closed by 8pm so we had an early night.

Hey Mister Tambourine Man

PETE:

The next day we drove on and it was quite distressing to see a sight that you don't get in the brochures. There are lots of kangaroos alright, but where we are they are all lying dead at the side of the road every odd kilometre or so.

Rule No. 1—don't drive at night if possible. The Roos just jump out any time and that is why all drivers have those big front double bumpers that reach over the car bonnet. It knocks the Roos for a six off the road and saves the car and I suppose the insurance. Poor Roos.

On we drive. Earlier we had bought 4 CDs for 2 bucks. There was some early reggae we had never heard of, Prince, Jeff Beck and a rather funny looking one called The Jesus Hokey Pokey, so we put that one on, then we took it off and then we put it on and shook it all about (sorry, couldn't resist that one).

However, we did put it on and Margaret loved it: a cross between hillbilly and gospel which meant she whooped and hollered all through it with the odd yee haa. We have still got that CD.

We come to a place I think we are gonna like. It's got a great name— Tamborine Mountain. We park just outside a rainforest entrance and there to greet us are our old friends the fruit bats. I am beginning to like these guys. In the morning, we go for a walk and discover that (here is one for the pub quiz) Tamborine Mountain has nothing to do with a tambourine or even Bob Dylan. It is an Aboriginal term from the Yugambeh Tribe meaning wild lime referring to the tribesmen who would eat the leaves from the trees to quench their thirst. I like the sound of that.

That is in one part of it and the other part we drove right up through the village up to the top of the rainforest to try and hear a Lyrebird and that, my friends, is nigh impossible. They can imitate anything and I mean anything, from chainsaws to mobile phones. You will hear them but you won't see them.

MAGGIE:

Let me tell you about my husband's dreams. They are so good that every morning I can't wait to ask him if he had any. Last night he had a strange one. Lionel Ritchie (I love him and Pete not so keen) visited him and he was singing a Peter Gabriel song to him. 'In Your Eyes' and doing a sort of Motown Era/Temptations version complete with dance moves. Pete's favourite two singer/songwriters are Peter Gabriel and Robert Plant. He just loves them: their lyrics in particular. Why Lionel Ritchie did a cover version in his dream we just can't quite decipher unless it's a birthday serenade to me. (Note to self—look up the lyrics of In Your Eyes)

It is another beautiful morning as per usual. I was sitting in the shade by 8.30 am. We headed up town to the charity shop and got some CDs for George. We get Prince; Jeff Beck greatest hits, obscure reggae and the Jesus Hokey Pokey. What a laugh we had driving down the road. It was so much better than the Church yesterday. Very uplifting. I was laughing like a kookaburra.

'You put your whole heart in, leave nothing out

you put your whole heart in and you shake it all about

The Jesus Hokey Pokey You'll turn your life around

He's what it's all about'.

We were thinking of following the road just by taking advice and talking to people. A helpful lady, who worked in Boonah supermarket, recommended the Lions Road but first we are heading to Tamborine Mountain which is nothing to do with musical instruments, Bob Dylan or the Mamas and Papas.

We parked for the night at the side of a rainforest area. One thing we notice is that we do not miss the TV one bit. We don't listen to the news, none of the incessant Brexit talk which is poisoning Britain. We have a TV in George, but we watch one DVD the whole trip—a Nelson Mandela film, *Goodbye Bafana*, and that's it. No more TV.

It is such a precious gift to give to yourself. Silence, the sounds of the natural world and the sound of your own heart.

PETE:

I have been bitten again, quite seriously and even my thumb has swollen up. I really need to get the insect lotion together. There is something missing from it and before I leave I make a note to find it.

We jumped in the van and drove on to our next stop wherever that is. On the way, we pass a rainforest retreat and wee white church. Quite amazing looking. We pass a guy who gives us the peace sign, "It's because we are in a flowery hippie van," I say and then I realise his fingers are actually facing the other way round. Well, he was quite young.

As Above So Below

MAGGIE:

We end up at a place called Canungra and the campsite is next to a creek. It looks a perfect place for platypus spotting. There are creeks everywhere in Australia. We look forward to seeing the names they have been given. We have had many a laugh. Wonglepong Creek; Little Ugly Creek; Bald Knob Creek.

As evening is about to descend Pete's bushman skills kick in. He is in his element making a fire and getting the dinner on. It is twilight and we see a wondrous sight: thousands and thousands of fruit bats flying out for the night, filling the sky for what seems ages.

There is a full moon tomorrow, which seems appropriate. Pete went for a walk and swears he turned round and something was following him, but he is not sure what. It had a long tail and it walked rather than hopped. We hear a rustle outside the van at 4 am. I look out and see a small creature with a wee kangarooish face. Was this what Pete saw earlier? Will have to investigate later. We decide to get up at 5am to see if we can see a platypus. No such luck but we do see fourteen large yellow crested cockatoos, two kingfishers, my kookaburra friends, grey and pink cockatoos and Ibis. We decided to stay another night...

PETE:

After a nice drive with no dead kangaroos to be seen, we see a sign for one of the showground sites our Aussie friend told us about. We drive down and it looks like paradise. We see the owner and I ask him,

Can we stay the night? YEP

Can we light a fire? YEP

Is there Platypuses? YEP

THAT WILL BE TEN BUCKS AND YOU WILL HAVE TO STOP TALKING SO MUCH IF YOU WANT TO SEE ANYTHING.

We settle down for the night with our fire on, just yards from the creek and sit quietly with a beer in our hands, waiting for our first platypus to show up. Then after a while we hear a noise and I know it is not a platypus. It's a fkn dog. A big, huge Motorhome has arrived with two dogs. Bang goes our platypus.

We decide to at least enjoy the fire. As we sit there we hear a low humming noise coming from above us. It got louder like a small plane or something. Then it sounds like big, slow, loud flaps of wings like the Nazgul in Lord of the Rings or a flock of swans. There it is, a huge black cloud just yards above us. It's them again—the Fruit Bats—only this time I remember seeing something similar on Life on Earth on TV. There are thousands of them. Around forty thousand or fifty thousand. They must be on a nightly mission to find fruit. It is a truly awesome sight. I wondered if our Christian friend, Priscilla, would be slavering at the mouth if she were here.

MAGGIE:

Next Day, while Pete was charging his phone, I went for a stroll along the creek to look for a platypus. A guy approached me and asked if I was the owner as he was making a film and this was a super location. He went on his way and as I ambled along I came to a beautiful spot with a rock perfect for sitting quietly to see what I could see.

At that I heard a rustle. A huge monitor lizard between three to four feet long from tip to tail, had made his presence known. I was relieved to see it was heading up a tree. I never really felt scared on our travels. The lizard was more scared of me I think.

I was sensible and stayed to the paths, never daring to go through long grass. I always remembered what our guide said on Heron Island—If you don't bother them they won't bother you. Australia gave me such a feeling of adventure and wonder at what we were going to see next. At that Pete came to find me and I pointed out the lizard on the tree.

We went a walk into town and got talking to a lovely lady in a quaint wee book shop. She recommended visiting O'Reilly Rainforest which turned out to be great advice. We asked her what she thought we had we seen last night. We wondered if it was a tree kangaroo but she advised that these are very rare and live way up north in the Townsville area. She reckoned I saw a Pademelon, a tiny, nocturnal Thylogale related to kangaroos or wallabies. She reckoned it was probably a possum following Pete.

I remembered it was a week since we got married and bought a bottle of wine to celebrate. Love Pete Burleigh. He makes me laugh and he has a good heart, which is hard to find these days.

PETE:

The next day we woke up to a truly glorious morning. After charging my phone, I decided to go a walk along the creek, keeping quiet in case of platypuses (or is that platypi), I sat at a rock for ages. I saw Margaret ahead talking to someone.

Later that day I saw the person again, only this time there was a bunch of them all pointing and talking. Turns out Margaret was talking to a film location Scout and what I was hearing was in fact the film being made. I heard something like "*As the little boy runs through the woods running for his life, his dog runs after him, giving him away. I'll be ready to get him.*" Since then I have been looking for a film called 'the dog in the wood' or something like that.

We make our fire again and settle down for our nightly Batfest and they don't let us down. It really is amazing to watch. After a while, the bats and the noise fades.

Just as we are putting the last log on the fire, I mention what a great night it is. Due to the full moon we can see for miles. Then I hear a different noise not from above, but from below. Here on the ground. They don't have wolf packs here, so no worries. The noise really does get louder and then I realise its frogs or toads—zillions of them. I sang a song to Margaret that I remembered from way back. Joy to the World by Three Dog Night.

We sat all night sipping our wine listening to all the Jeremiahs. Another weird and wonderful, noisy experience.

MAGGIE:

The sounds over here are something I will never, ever forget: kookaburra's laughing, wooo wooo birds and their ghostly love ritual, thousands of grasshoppers sounding like alien transmitters; butcher birds with a call sounding like the wee tune from Close Encounters; rainbow lorikeets shouting and squawking at each other; fruit bats and the melodic flapping of their big wings; cockatoos screeching loudly and mischievously.

Last night it was the full moon, It was a Supermoon and reminded me of a huge lamp. We have never seen anything like it. Huge. Most of the time most of us seem oblivious to the powerful effect the moon has on every single living thing on this planet.

There was a strange noise like a machine gun/motorbike type of noise. It could be frogs we think, but there must be millions of them as it is so noisy?

I asked the next day and was told it would probably be the Cane Toads. There must have been thousands of them all looking for a mate. I got the distinct feeling there were not enough girls to go round.

We were horrified to find out that these toads are not native to Australia and are causing devastation to native species. They were introduced by sugar cane farmers to control destructive beetles, but they have a deadly poison secreted from their skin which kills any predators. Another example of humans upsetting the balance of nature once again.

PETE:

Back in the van we head to a place called O'Reilly Rainforest. It has a Rainforest Treetop Walk. We look around to see where we are going and come across a bunch of people all handling birds. It cost 5 dollars to buy the seeds, then the birds like to come and eat out your hands. I am not going to bother with that as I like to think they will land on me naturally. They look like parrots but I think they are parakeets: green, red and orange and other colours as well.

As I was standing, a girl came up and trickled some seeds into my hand the way you trickle sand through a clenched hand. The birds came instantly. One landed on my shoulder I immediately bent my leg and said, "Aaaaaar me hearties." No one laughed. I think it was because they were all Chinese and had never heard of Long John Silver.

We moved onto the treetop walk along the canopy of the forest. It is a swing bridge. Only a certain amount of people are allowed on at a time in case it breaks and you fall all the way down to the ground and probably die and get eaten. It is scary enough walking on a swing bridge without getting told you could fall…and it is even more scary when loads of Chinese people, who don't seem to read English, all come on to the bridge after you—all taking selfies.

I don't know why, swing bridges don't bother Margaret, Maybe because she was brought up on a farm? Finally, they all stop taking selfies and we all get off the bridge safely. We walk round a nice garden and a secure walkway capable of taking wheelchairs. It is nice that some places can do that and let some less fortunate people do rainforesty things as well. A last look at the birds and we drive on.

It is a drive up a mountain road and takes quite a while. We stop at a fairly large area of flat ground for a ginger beer. As we are standing looking at the beautiful views a girl walks past us. She is off the tourist bus which has just arrived with the same Chinese tourists we had met at O'Reillys. Margaret points out that it's the same girl that gave me the bird seed and we end up talking. She comes from Chicago, is travelling by herself and she is very cute. Margaret gets her pic with her then to top it off a couple of kangaroos hop past and with that we wave goodbye.

As we are travelling though real Australian scrubland in our wee van it is a real feeling of freedom. We both agree that this is great. Then we see a head popping up over the top of the grass, then another, then another. Jings! There's loads of them and they've all started bounding away at the one time. Real Aussie stuff. This, I think, is called a Troop of Kangaroos or a Court. Whatever it is, they are well away, bounding higher than Zebedee.

MAGGIE:

When we arrived O'Reilly Rainforest, we noticed a Buddhist type mountain garden. Apart from a young girl walking serenely stopping to look and smell plants, we are the only other people. We savour the moment.

We then had a great birds eye view from the treetop walk and there are lots of birds up here. Later, we noticed a group of people with birds on their heads, shoulders and arms and notice that they are feeding them. Mostly bright parakeets and zebra finches.

The young girl we saw earlier gave Pete some of her seeds and immediately a parakeet landed on his shoulder and he did his Long John Silver pirate impersonation. Then lots of birds landed on us. It was a joyful moment.

We drove further up the mountain and came to a stopping place with a stunning panoramic view. We sat and soaked it and the sun up for a while. The tour bus we had seen earlier pulled up and lo and behold the same young girl came off the bus. I said to Pete that is three times we have seen her and sure enough she ambled over to speak. This delightful girl came from Chicago and had been travelling by herself for some time. We took pictures and exchanged pleasantries. The bus tour guide started to speak and so we listened in. Turns out the beautiful view is indeed of Tamborine Mountain.

It is great having George the campervan, stopping where and when we like. We see our first troop of kangaroos in scrubland. At first, it looked like maybe five of them but when they spotted us lots more heads popped up and there were maybe around twenty and en masse they hopped ridiculously up the field. Such strong and fast animals. I have the utmost respect for them as they have been persecuted for centuries. No wonder they don't like hanging out with humans. Who can blame them?

We keep stopping at creeks. We sat for another couple of hours at Canungra Creek, our personal favourite. What a treat. Still no platypus though.

Jings, Crivens, Help Ma Boab

PETE:
Earlier in the week at Boonah, a woman told Margaret that it's a must that you drive down the Lions Road. It is well off the tourist track and well worth it she said. So Margaret decided that since I look like a lion that I should drive so that I can say I have driven the famous Lions Road. OK no problemo. Looking forward to it.

We fuelled up at a place called Rathdowney, had some local farmers honey complete with honeycomb for our breakfast. We popped into a quaint museum where there was a certificate on the wall admitting Arthur Conan Doyle (who in case you don't know wrote Sherlock Holmes) into the Oddfellows Fellowship. Methinks I could get admitted quite easily. We swap drivers seats and away we go. Lions Road here we come. TOOT TOOT, HONK, HONK.

JINGS: I wasn't expecting that. It's like driving round a helter-skelter. Too late now, we can't stop. The road is too narrow and something could come round the corner. I feel like I am going in circles, with all these turns my foot is always on the brake.

CRIVENS: There's a rickety, hickety, wooden slatted bridge right in front of me. It looks like something out of Indiana Jones: this is real squeaky bum stuff. I manage to stop in time. "What now?" I say.

"We need to cross," says Margaret, "we can't go back up the helter-skelter."

I jump out and go down the banking and look under the bridge. I don't even see any supports, but I do see lots of beams of light shining through the slats. This bridge is unsafe, I decide.

So I get back up the banking to the van and there is Margaret at the other side waving. "IT'S SAFE," she's shouting. "COME ON, YOU'LL BE FINE."

I shout back, "I'M IN A VAN, YOU JUST TIPTOED OVER!"

"YEAH, YEAH," she shouts. "WE'RE INSURED. C'MON."

I don't pray that often, but I say sorry and decide to try say a quick one. All of a sudden a motorbike screeches to a halt, the guy looks at me, does that thing my biker pal does: fist in the air he revs up as if it is a death wish or something and ZOOM off he goes. Plenty of noise but over safely. I feel as if I have been given divine encouragement, so with my heart going BOOM BOOM BOOM, I do a big shout out, do a mad Haka face and put my foot down.

HELP MA BOAB: I forgot I had four wheels and I feel like I am on a rodeo horse or even one of those mad bulls that fling you all over the place. I look in the mirror just as my head hits the ceiling. I could picture lumps of wood in the air behind me. Finally, I am off the bridge, just missing a laughing Margaret. "There, that wasn't so bad, was it?" I make a mental note, it's called KARMA.

MAGGIE:

Heading to Rathdowney, the satnav says it will take forty minutes. It takes over two hours due to stops (Roos, Platypus searching and beautiful sunset). Going to do the Lions Road tomorrow. We are heading to Kyogle and Fairy Mountain.

We parked in the pub car park and had a couple of Schooners of ale before having a very good sleep.

Feel like the Queen of Hearts as we have bread and honey/honeycomb for breakfast. Glenapp Station had an excellent map of the Scenic Rim. We have some real good memories.

We nip in to see the Museum before we head down the Lions Road. There are some interesting bits and pieces and I am taken with a poster about good manners. It makes me think everywhere we have been so far we have been treated so well. The good humour and nice manners of the young people in particular is so sweet. We often get greeted with "How you going?" I suppose the Scottish equivalent of HOWZITGAUN. With a warm smile and genuine interest in the question.

The Lions Road was harum-scarum in the wee van but I reckon in any type of vehicle. You need a lion heart to survive the single track road and all the numerous bridges and creeks. One bridge was a bridge too far. An ancient swinging bridge. Luckily we survived to tell the tale. Well done Pete the lion heart.

We don't really connect with Kyogle. We decide to keep on moving on. Firstly though, we nip into the Information Centre which are always worth a visit. Often run by volunteers, they have loads of leaflets but it is the volunteers we especially like speaking to. They are full of interesting facts. This one turns out to be an elderly gentleman originally from Canungra. We asked about the platypus there and he said, "yeah, mate, they are everywhere. They come onto the grass sometimes." I feel like they are like wee cartoon characters, who have been dancing behind our backs dressed in hula hula skirts and then hiding when we turn round lol.

PETE:

It's back to Margaret driving. This time it is a nice n' easy drive and we come across a sign that says Nimbin. Being a Hippie I remind Margaret that this is the Hippie capital of Australia. Although I gave up smoking and all that stuff years ago, it will be well worth a visit as it should have a real, peaceful, groovy vibe and our hippie van will be right at home.

WRONG: what a huge disappointment. No flower power, no sixties style groovy music, not even Bob Marley or the Beatles. In fact, it was so touristy I think I was the only genuine, real hippie there. It's all leave our children alone and let them come in when they want stuff and give them anything they ask for so as not to hurt their little feelings. Stop killing ducks for duct tape. Posters with stop killing mother earth our little feral children need her. We decided to move on after a short while. Thank God this is the only place I haven't liked since we got here and I thought it would be one of the best...

MAGGIE:

We went to Nimbin to see the so called alternative living. It had an uneasy vibe and is very, very touristy. The shops all selling the same hippie stuff. The place has an air of neglect and the people don't seem to be as friendly here.

I was interested to hear the story of how Nimbin came about. Seemingly way back in the early seventies, around Woodstock era, a group of free thinking hippie types decided to set up a community for alternative lifestyle: growing and producing their own food etc. Sounds ideal. However, they decided to let their children be 'free' to do anything they wanted. Unfortunately the children ended up unhappy with their freedom and turned to harder drugs than their parents. Everyone needs to have boundaries. This is the third generation and the lifestyle now shows up as delinquency. The town is not thriving. I was very happy to leave.

PETE:

Here we are in Lismore. It's quite a big place. We decide to stay the night as it is late. I have been bitten again. It's a daily thing. There is still something missing from my insect lotion, but at least it is helping the bites a wee bit.

We went to a couple of Churches. The first one was full of old folk. I noticed they were wearing jeans. My dad would never be seen dead in a pair of jeans. We had sandwiches and a coffee and then head down the road into the main town area.

I see quite a crowd going in and out of a building. We head over to hear the end of a tune and to the sound of people clapping. I wave Margaret over and in we go. The band started up and played a happy clappy song. I turn round to see Margaret with her hands in the air clapping with the rest with a big grin on her face. Looks like we are here for the day.

Afterwards, we talk to a guy who actually says FAIR DINKUM. He says even the Australians reckon its harum-scarum to drive the Lions Road. So it's a kind of achievement and I feel good. In fact, if the next charity shop has a lion medal or something I might just get one and wear it.

MAGGIE:

We move on and its quite late. We end up in Lismore. I notice there is a Church. There is service at 8.30 the next morning. I set the alarm. It is very much old school with an elderly congregation and it's all about our 'sins'. The hymns were a bit dreary and I didn't feel like singing them. The sermon was quite good. The Preacher was a kindly looking, elderly man wearing a check shirt and jeans. He told a few jokes one about Footless Fred—the goat and the window 'mirror'. My favourite story about a wee boy who broke a favourite toy ornament and was crying. The dad said don't cry, but the mum said sit on my lap and we will cry together and then we will remember how nice it was and see if we can fix it or replace it. I like that message of be true to our emotions. There were lots of quotes emphasising our sins. They were nice enough, but we left quickly as this Church is not for us. On the way out, we were told we were welcome to come back as The Devil is everywhere.

Pete took us up a side street and lo and behold we hear music, beautiful music, which drew us in. Very busy: full of families and all nationalities. It was a breath of fresh air compared to the other church service. JOYOUS. The Preacher spoke about Divine Intervention and help being everywhere if you ask and listen. He quotes from the Bible about belief in self and hope. Set goals ask for divine intervention and then put it into practice to accomplish the goals.

The band played a beautiful anthem and we all sang about Spirit blowing through us. It felt Native American or Māori. After the service, we were invited to have a delicious Barista coffee and a guy came over to talk to us. A cool dude. He is full of joy and non-judgemental. "Everyone finds their own way," he says. Speaking of ways Pete tells him about the Lions Road "FAIR DINKUM," he says, "Mate, that road is a nightmare. Even the Australians think it is harum-scarum." So glad we stayed in Lismore.

PETE:

We drive in the glorious sunshine to Byron Bay. I've heard of this on the TV back home and it is real Australian brochure stuff: golden beaches, blue skies, loads of bikini clad wiggly women for the men and plenty of surfboard dudes for the women. Everybody's happy. It's not long before Margaret spots an ice cream shop. "They look the bizz," she says, "you need to get me one please."

"Erm OK," trying to work out what six dollars is in our money. Turns out you get double scoops and they are very, very tastie with no wastie. Man, they are delicious. So glad I convinced Margaret to have one.

We head back to where we are camped. It is a place called Suffolk Park. It is a real nice place with a good vibe and we decide to book in for three days, rest ourselves, put plenty of charge in the battery, wash clothes, shower etc. Give ourselves and the van a good rest and airing. Next morning we wake up early as everybody does. It is another glorious day. I pick up one of our CDs Jeff Beck so here I am listening to Hi Ho Silver Lining at good volume, in Australia, sun shining, in a hippie flower power campervan. Love eet. Then another hippie campervan pulls up across from us. It is the same as ours. We nod a kind of brothers-in-arms nod to each other and enjoy the moment.

MAGGIE:

Found a campsite at Suffolk Park which will allow us to walk into Byron Bay as it is a nightmare to park in. We booked for three nights to charge our batteries and George's battery. We notice when driving George that he does not like to do more than eighty kms per hour or he starts to shake and rattle and we are really hoping he doesn't roll. We are happy to take our time.

Did the washing up listening to Jeff Beck CD with Rod Stewart on most of the vocals. Music is great and it really is Rod at his best.

We enjoyed the walk into Byron Bay and realise it is about seven kms. We wandered around the shops and saw a guy playing a banjo. Sit and listen. We go up and speak to him and say he reminds of us Billy Connolly which really pleased him. Ends up he lived in Newcastle UK for a few years. Enjoyed our blether about life in the UK and life in Australia. I notice that there are loads of buskers in Byron Bay.

Had an amazing ice cream one scoop lemon cheesecake and one scoop Belgian chocolate. We wander along the beach and notice there are lots of surfing dudes and dudettes and enjoy just watching them do their thing.

Pete buys me a beautiful set of Nature Spirit cards by Denise Lin. I absolutely love them. Whilst waiting at the bus stop to go back to the site we saw and heard hundreds of cockatoos roosting in nearby trees. Ends up the last bus is at six so we missed it. We get a taxi and the driver recommends that we try to visit the seven mile beach at Lennox Heads and also the Whian Whian falls further up the road.

There is another identical Hippie Camper van right next to us with the name Ray. We spoke to the young couple and they were saying they were like us finding it difficult to get to grips with him hopping all over the road.

PETE:

We decide to go into Byron Bay again and do a bit of shopping and touristy stuff. In one of the shops, we learn that they guy behind the counter used to live in Lockerbie, Scotland.

We pass a hotel on the way back with a notice for a 'rump steak dinner'. It reminds us of our wedding day so we decide to put on our glad rags, give ourselves a pamper and book a table for two. As usual it is great. As they say in Australia: 'you can't beat a good rump, mate'.

MAGGIE:

We had a long lie in till 8 am and get the bus into Byron Bay. The driver is very nice and says he wants to visit Scotland. Lots of Aussies have been to the UK and there are Scottish names everywhere over here. A lady whispers to me that its free bus day on Thursday to encourage more use of public transport. What a great idea.

A guy in a shop got all misty eyed when he heard our accents and gave us a fifteen dollar discount. So nice to talk to him and remember the nice things about our Country. His wife is a Geologist and that's why they lived in Lockerbie. He just loves our Scottishness.

We are going out for a rump steak dinner at the local hotel. We wore our wedding clothes and had a rump and a pint just like we did on our big day.

PETE:

We liked Byron Bay, but as we are travelling grey nomads we move on. This travelling is magic. Anybody reading PLEASE DO IT. Eventually, we pass over a very large bridge and there in front of us, is a giant prawn made out of corrugated iron. It is huge.

MAGGIE:

It feels the right time to move on. We notice there are a lot of residential caravans here. Great campsite but we love the freedom of the road and not knowing where we are going to end up. We nip into Byron Bay one last time and get a lemon cheesecake and espresso ice cream for the road. It was raining quite hard. The first rain for over three weeks. We end up at Lennox Heads without trying. We see the gorgeous seven mile beach which has to be seen to be believed.

We got a sharp knock on the van at 7 the next morning and a yellow card for illegal parking near the lakeside. We apologise. He says we will get fined if we get caught unauthorised parking again in the district.

We park legally and go for a lovely walk along the seven mile beach. We see a sign beware of snakes and an even bigger sign saying this is a world shark hot spot. We look around and there are lots of folk boogie boarding and surfing in the sea. People just get on living their lives. There are a couple of coastguards on hand.

Later on we notice a newspaper with an article about a guy who had half of his thigh bitten off by a shark last week. He said that after his plastic surgery and rehabilitation he would be straight back in because a shark attack is rare.

A guy at the surfers cafe says that heaps of Scottish people have settled here in recent years. The seven mile beach with its constant charge of white horses might have something to do with it.

It has been a difficult day for us today. The bites are getting to Pete. He has been very quiet. He had his Journey T shirt on with the mirror reflection of a beautiful lake and forest on it. It feels appropriate. I remember me finding it for Pete when we were just great friends. We feel it has come to symbolise not only our love of travel, but our relationship. We are like mirrors where we reflect and learn from each other. Each and every day is a school day.

As we drive along I can't believe my eyes. It is a giant metal sculpture of a Prawn.

PETE:

We drive to a town called McLean. I immediately think of my pal whose name is McLean. As we drive through it we discover it is totally Scottish themed, even the telegraph poles are painted in different clan tartans. We decide to find a campsite as we like the feel of this place.

We talk to different people who tell us all sorts of things. The most helpful was a wee lady in the Information Centre. She knows our accent straight away. We end up telling her we got married at Burleigh Heads she immediately says congratulations and gives us a thistle fridge magnet that she made herself.

McLean holds a Highland Games every year with pipe bands. Every street, more or less has a Scottish name like Ayr Road and Glasgow Street. We see a fish and chip shop. It sells that fish from the diner (remember Sheila of the towelies). The fish is called Barramundi so I have that and Margaret spots prawns. They are not the size of the one we passed, but they are very large and very fresh.

MAGGIE:

We arrive at a quirky town called McLean. So proud of its Scottish name and connection; it seems more Scottish than most Scottish towns will ever be. There are two hundred and forty different tartans all over town painted on telegraph poles, a cairn and Scottish themed pictures everywhere. The butcher sells haggis and has an 'up yir kilt' sticker in his window.

We visited the Information Centre and a lovely lady in a kilt greeted us. Like most Australians we speak to, she had been in Scotland. Turns out her granny came from Montrose and she is proud to tell us has a telegraph pole with her tartan. She gave us lots of great information re local sights and when we told her we got married at Burleigh Heads she beamed at us as she gave us a thistle fridge magnet, she said she made herself, as a wedding present. Thank you lovely lady.

I went to a baker and bought a scroll and asked for a spoon. She said all she had was a spork. The very thing, its a cross between a fork and a spoon. The scroll looked better than it tasted. It was very stodgy with apple, cinnamon, sultanas inside and icing on top. Very heavy, but I persevered.

We find a Showground next to the Clarence River and are going to stay for three nights. There are lots of grey nomads in here.

We spoke to a lady at the showground. She told us the whole building got destroyed twelve months ago in a cyclone and is just re-opening this weekend. She saw the roof actually get blown off. There is no sign of whinging. Just a good humoured resilience.

We sit down by the river and Pete feels like doing one of my Nature Cards. He shuffles the cards and one flies out of the pack. When he picks it up, it is Flowing River card. We reckon we are at the right place at the right time. Just at that we hear a familiar noise and we look up to see thousands and thousands of Flying Foxes swishing above us. We feel so blessed to see them again and shout woo hoo and wave.

PETE:

We find a the Showground campsite. We meet a lady who tells us that this place got hit by a cyclone last year and they are rebuilding it as no-one wants to leave McLean or their homes. We sit down next to the huge river which we have found out is called the Clarence River. It is so wide in places it's as wide as a lake and it goes on for miles and miles. All of a sudden we hear a familiar noise getting louder and louder, the sky is getting darker and we whoop. It's them again. it's our old pals the fruit bats. We wave to them like a pair of idiots.

61

MAGGIE:

We went up to the viewpoint recommended by our lovely Information lady and had our breakfast looking over the hinterlands, lakes and of course, the amazing Clarence River. The Aborigines have a story about The Great Serpent and how the different tribes from here got their names. From here, we can see the river winds its way as far as the eye can see.

We see an eagle as we are speaking to a lady originally from Belfast. She came over by ship in 1964 using a £10 ticket. She had a wee cairn terrier with her. As I finish my tea I look up to see three kookaburras sitting on a branch. I cried to Pete to look up and just at that the three of them put their heads back in unison and did their thing. You just couldn't make that up. Priceless. Love it. Then another one landed on a branch next to me.

When we came back down, the Showground was full of horses practising for the upcoming Highland Games. We spent a while just watching them. We heard the sound of bagpipes. It felt surreal. The pipe band are playing Scotland the Brave. We are going in search of the Kerr tartan. It is my mother's maiden name. We eventually find it across from the chip shop. The prawns are the most delicious prawns I have ever tasted. No wonder they call Australia the land of plenty.

PETE:

I wake up to discover that I am bitten all over again. I need to find this missing ingredient for my cream and quick. It's a lovely day and we decide to go to the viewpoint for breakfast. As we have a cuppa Margaret points out, on a branch, three of her pals the kookaburras. They all put their heads back at the same time and Kookabrood. I think she would have jumped up and joined them if I wasn't there.

There are a bunch of horses in the showground and young people are learning show jumping, dressage and fast riding skills. We watch that for a while and then head round all the streets. Margaret wants to find the Kerr tartan as that is her mum's maiden name. It takes ages but it is good fun, some of the streets are even in Gaelic. We finally find it and its straight across from the chippy. Nice one, I go for the prawns this time.

MAGGIE:

We spend the day doing the laundry, airing the van, hanging the washing out. Feels really good. We are watching the young folk and their horses again today. We got talking to a Canadian lady who is a metal detectorist and among other things tells us she has found florins and a medal tracing back to a settler before Adelaide was formed. She gives us tips where to look. Pete fancies a metal detector now.

She married an Australian mining engineer and they made the conscious decision not to have children. They have been all over Africa, America, South America, New Zealand. It was great talking to her about her travels. She says Australia is tame compared to Canada because Canada has bears, mountain lions and wolves. She told us one day she was out a walk with her Italian Mountain dog, when she encountered a bear and she really thought she was a goner. She looked at her dog to say sorry and I love you and then she looked the bear in the eyes (not recommended). She doesn't know how, but she managed to pull her dog and herself into a toilet and shut the door and just sat there for hours with her legs and feet up against the door waiting on the bear rampaging in. Much later someone came to use the toilet and it was only then she knew the bear had gone.

I could feel how that was REALLY looking death in the eyes. She also said she nearly stood on a snake and it didn't attack her, so she feels very, very lucky to be alive. She recommends that we visit Newfoundland and says the locals still have Scottish accents. Sounds a great idea to me.

PETE:

We are sipping our ginger beer when we met a Canadian woman walking past. Her husband is a mining engineer and they have travelled the world, moving every four years or so. We found out that Australia fits into Canada twice. Imagine that. I am gobsmacked I didn't realise Canada was that big and that Britain fits into just one of its states four times. I told her I am getting bitten every day by mozzies and she says I am lucky because in Canada everything wants to eat you AND they chase you. Bears, cougars, wolves, moose, buffalo. She also says to stop opening the van boot the way I do with my hand under the ledge as that is where poisonous red backed spiders hang out especially wheely bin ledges. We say goodbye as she likes to walk around with her metal detector.

We sit back down again, see a pelican, an egret and then, as if they know it is our last night, here they are: The Bats. The most I have ever seen. It takes nearly half an hour for them to pass. Again there are thousands and thousands of them. What a sight seeing them flying over this huge river above us. I sure hope to see them again.

MAGGIE:

We had spotted a Presbyterian Church earlier and were told there is a service at 6 pm. I feel I would rather stay to see the sunset and of course, our bat friends. I said to Pete as I looked around at all the beauty. This riverside is my Church and Pete said he had just thought the same thing a minute earlier. We feel blessed.

An Egret came right up to Pete and then to me. A young family came to investigate and we got talking about birds and wildlife. Turns out she is a Jehovah's Witness. She doesn't try to convert us. She is lovely. We have an interesting conversation. She says she believes that going into nature is very spiritual. Confirmation of our earlier thoughts.

We are on our way to Grafton—tree city of Australia. The Jacaranda are not in bloom just now, so we only stay a wee while. Time to move on in search of platypus. Didn't do enough research as the 'town' we come to has about four houses, so we move on. Then we realise that we are in the middle of nowhere and running low on petrol. Oh dear, I am trying not to panic.

We see a guy driving down a dirt road in his truck and Pete waves him down and asks him where the nearest petrol station is. He says the nearest one is about twenty five miles back the way we came. OK we say and turn round and start to drive back. Just at that he comes back over the bridge, flagging us down this time. We stop. "Hey, I've got a twenty-litre container of petrol you can have." He has got a kind, friendly, honest face. "It's yours for twenty bucks." DEAL we both say together! Another example of how kind the Aussies are. We thank him very much and are on our way again.

It takes ages and ages and lots of kms to get to the next town. We thank our lucky stars again re our Aussie knight in shining armour and then see a sign for a place called Dolingo.

We stop at a small village to ask the way and I could feel an uneasiness come over me. I can only describe the guy as an authentic hillbilly. Not sure if the woman was his sister. They were staring at us in a funny sort of way and then a big, fierce looking, but at the same time, stupid looking dog came out and we decide to leave quickly. That was confirmation that dogs look like their owners. We are so glad to see houses in the distance. We were tired as we have been driving for hours and we manage to park for the night.

PETE:

We are passing through what seems like Hillbilly country. There is no sign of any towns and we notice that our fuel is going down. We stop just over a bridge as we are not sure what to do as God knows how much petrol we've got. After a while, I see a pickup truck coming down a dirt track, he turns and comes this way. I put on my Bush Kangaroo Hat, try and look non touristy and wave him down. "G'day, mate. Where is the nearest petrol station?" He is not taken in by my hat, but he looks an honest enough guy and says, "About twenty five miles up the road but it shuts at 5 and it is 4.30 ish" and off he goes.

We decide we need to go back and turn to go back over the bridge. At that a pickup truck drives towards us, waving us down. It is the same guy. He says follow me and I will get some from my farm. I realised that it might be shut by the time you get there. Where's the farm I wonder and is he part of a Ned Kelly type gang? It's just at the corner we have already passed it. Turns out he has a drum that is lying doing nothing at his farm. He fills up the van and we give him twenty bucks. Everybody is happy.

As we drive on for miles we eventually come to a hillbilly place and as we feel lost we stop. G'day I say is there a camping ground about here? Honest the woman just stood there toothless, a dog barked and it looked toothless as well. She is still just standing there staring, then a man, who looks like a mixture of Jed Clampett and Worzel Gummidge comes out and the three of them just stand there staring. Then they all grin, even the dog and we don't wait for their neighbours to come out. I swear we could hear dualling banjos as we left.

MAGGIE:

The town looks better in the morning light. We went to the Visitor Centre and once again got lots of useful information. The lady there recommends going to see Danger Falls. She also confirmed that indeed she had seen a platypus but, that they are very, very shy.

She told us there is also a huge collection of trains. We go to see the trains first and think that it could be turned into some sort of camping experience. We then head for Danger Falls. It is such a beautiful view, so we decide to go for a walk down to see the falls at ground level. They are pronounced Daan ger and not Danger so Pete goes in for a swim as there are a few people in swimming.

When we come back up, we notice that there is a Labyrinth and a notice about the circle of life. We decide to walk around and I really enjoyed the experience. Really slowing down, contemplating on how far I have come and how I am still learning, step by step, every day, to enjoy the moment.

PETE:

We went into the Tourist Information Centre at Dolingo and they recommended Danger Falls pronounced DAANGER. We visit a place that is full of trains and they still look good and it looks like they could be used for holiday homes as each of the carriages are painted in bright colours. I think they look real cool.

Away we go to visit Danger Falls. We park the van and after taking in the view we start to walk all the way down. It is a wooden pathway with trees growing up through the middle. It's sort of U shaped so it is a welcome surprise when you eventually get there and it is just what you imagine. A kind of oasis. There's folk in swimming in it. I decide to go in as well. It is a welcome relief in the water for me. My skin really itches all the time with these bites. I keep rubbing stuff on, but there is something needed in it yet. We come out of the falls.

When we get to the top of the steps, we see a Labyrinth on the ground. I've seen one before. The idea is that you walk your own path, stopping and starting at will, till you reach the centre, then, you walk back out and your life is no longer a maze. Sorry I just made that up, but it is something along those lines.

Once we left the labyrinth, Margaret got bitten by a flying ant. "Welcome to my world," I say.

MAGGIE:

As we made our way down off the plateau we realise just how high up we are. The roads are very steep and winding our way down we stop at a lay-by to enjoy the beautiful view. There we meet an Australian couple who enjoy cycling. Pete was telling them about his bites when she told him she had been bitten by a spider whilst digging up sweet potatoes in the garden without wearing gloves. Her finger was very swollen but she was a trooper. True to form of the Aussies she is very pragmatic and generous. When she produces a spray and some anti histamine tablets and give them to Pete, he is most grateful. I have no doubt she will be back out, digging her sweet potatoes, when she gets home, but this time with gloves on.

She tells us about her son, who now lives in Chile. When he just left school, he did a gap year exchange in Iceland and has never looked back since. His friend got him four platinum Pink Floyd tickets and he insisted he took his mum and boy was she glad she didn't babysit. She said it was an experience she will never forget.

As we were leaving her husband said I notice you have a Mitsubishi campervan. I used to have one and called it Shitsubishi.

PETE:

We stop at a small town called Bellingen and I get some histamines from a Pharmacist. My itch is bad and so are my bites. The insects really like me and they don't bother Margaret much.

As I walk up the street there is a giant map on the wall. Looking at it I see the name Taylor and someone has written next to it 'the pub with no beer'. My grandson is called Taylor so we decide to go and get a photo of the signpost to send to him.

We arrive and sure enough there is a big sign Welcome to Taylors Arms, The Pub with No Beer. We decide to go in and it is the busiest pub we have been in yet. To our relief they are all drinking beer. We just get in and the landlord asks us to pull a raffle ticket and someone wins a meat pack full of steaks.

We investigate the name. Way back during the gold rush the story goes that a guy was heading home and it was taking days and days to get there and the only thing that kept him going was the thought of a nice cold beer. However, when he reached the Taylors Arms all the Americans had drunk the pub dry. He was that mad and angry that he wrote a song about it. Years later, Australian folk singer Slim Dusty got a hold of it and it reached the Top 20 in 1970 something. Why not google Slim Dusty—The Pub with No Beer and hear it for yourself.

MAGGIE:

I am realising that these insect bites are affecting Pete quite badly. He is very quiet. We end up at Bellingen showground and then head straight to the Pharmacy for some anti histamines. We then relax by the riverside and I read some of the book I got from the Aborigine shop in Byron Bay. Secrets of

Aboriginal Healing: A Physicist's Journey with a Remote Australian Tribe by Gary Holt. We both fall asleep in the sun listening to loads of young folk enjoying themselves.

That night we see our bat friends again and think there must be a lot of fruit in this area probably because of the river. They are very good to us humans as they spread fruit seeds all over the place.

Had a long sleep. Don't get up till 8.30 am. By the time we have breakfast and shower, the campsite guy was looking at his watch. Time to be on our way. We stop at a cafe and have a banana and fig loaf with caramel. I have never had fresh figs before. No wonder the bats like them.

It is sweltering hot and we decide to drive on and end up at Macksville. Pete saw a sign for Taylors Arm and the Pub with no beer. We decide to investigate and drive up the pretty valley. We see a very busy pub with a humongous log outside it. Seemingly during the logging days, it was so heavy they couldn't shift it and so now it is a sort of monument to a bygone era. The pub is too. It is full of old mining lanterns, old saddles and suchlike. The landlord asks how long we are staying as the first prize in the raffle is a schooner of ale a week for twenty weeks. We draw a ticket and someone wins a meat pack. We see a poster of Slim Dusty who was a huge star around Australia. We have a schooner of ale each and it's time for bed.

PETE:

The field across from the pub is the campsite for five dollars a night. We can see it is dark and gloomy looking. Suddenly we see lots of lights flashing in the distance. I remember seeing a poster for a Phil Collins concert and wonder if it is an open air concert. Just as I was about to say to Margaret to go and see what it is, we realise, to our horror, it is our first electrical storm. It is a belter. Each flash lights up brighter than daylight and then we hear the thunder claps. As long as the lightning is not forked and not hitting the ground we are OK I say. Just at that the biggest fork lightning I have ever seen hits what looks like the next field away. We lie next to each other during the storm, every now and then looking out at the lightning. It's scary but hypnotic.

MAGGIE:

We just saw the biggest lightning display in an electrical storm. People would pay money to see this sort of show. It lit up the sky for ages with no breaks and

I thought OMG what will this be like when it reaches us because, at this point, there was no thunder. It brought out childhood fears as my mum's sister's house got hit by a thunderbolt when we were young and blew all her electrical plugs, TV etc. We used to hide under the big kitchen table when we were small. I thought I might as well surrender to my fears. I could feel my heart racing. The thunder was getting louder and nearer. If I was going to go, I was going to go. It went on for at least two hours. Seemingly this is normal. When it got nearer, it wasn't as bad as I feared. In fact by then, we were enjoying the show. Bring it on.

In the morning, the air was clear. We could see some amazing big, black cockatoos with yellow markings flying over to the nearby trees and of course, six kookaburras.

Pete was talking to a lady who (surprise surprise) had been to Scotland. She got chastened by a policeman in Dunfermline for going up a one way street the wrong way. Pete says he saw her go down on her knees. He thought at first to pray, but it was to wash her hair and face at the rainwater container. Wasn't that a great storm she says. Just what we need!.

PETE:
Everything is clear in the morning and we see five or six great, big black cockatoos eating pine cones. They have yellow throats and I think they are quite rare.

We are heading to Port McQuarrie. We found out there is a Koala Hospital there. They bring in injured koalas from all over New South Wales and even Brisbane. They are very slow moving and are quite often hit at night by traffic. As we were there three koalas are brought in boxes. We manage to watch through the glass as the vet examined them. We manage to get a guided tour and see lots of them with the volunteer handlers. The hospital doesn't receive funding so all donations are welcome. It is sad to see them hurt. They are so adorable looking. Everybody knows they eat Eucalyptus leaves, but did you know they are actually called Gum trees. They like the blue gum leaves and are very fond of the red gum leaves, but their all-time favourite is a non-native tree from Yorkshire called the Eeebahgum tree.

MAGGIE:

The skies are Scottish grey and much cooler after the burning heat of yesterday. We are heading to a Koala Hospital at Port McQuarrie. The roads are nice and quiet. We arrive around 2 pm and go to the Roto House Museum first of all. The lady caretaker was very welcoming and when I said she reminded me of my auntie Agnes she took it as a huge compliment as once again she has Scottish connections. One of her ancestors was a Welsh convict who fathered two children before heading to the gold rush in San Francisco. I wouldn't be here if it weren't for him she declares with a smile.

I enjoyed looking at the old photos. The house was bequeathed by a family of Irish/Scottish descent. They had six children; two girls and four boys. Three of the boys went to war and came back, but all were affected by the war and the only person who married was the boy who stayed at home to farm during the war. His wife died at twenty eight after giving birth to a daughter. That daughter became a nurse and had nine children. Many descendants live here and the rest, as they say, is history or herstory.

Got to see koalas up close. The tour was informative. Koalas are not bears they are marsupials. Some of them are permanent residents because they could not survive in the wild, but the aim is to nurture as many as possible back to the wild. Three more were brought in and we saw one lift its head out of the box and put its ears up just like the one we were privileged to see in its natural environment at Burleigh Heads.

PETE:

We leave the Koalas and head to Gloucester. It is recommend we visit the Billabong gardens and walk round taking in the Aborigine names of the plants and trees. We notice a market on nearby and have an ice cream and freshly squeezed orange juice.

There is a large mosaic platypus which we rub for good luck. We head to the Barrington campsite and get settled for the night. Have our lamb steaks.

Next morning we are off to Stroud to another showground. We see a couple of camels on the way. Have you ever seen a camel with its long neck stretched out to graze on the grass? It is funny looking. I bet if they kept their necks straight when they are upright they would be bigger than a wee giraffe.

In Stroud, once camped for the night, I opened the van door and saw a possum right in front of me, climbing up a tree. It stayed there for a good wee while and I get a good look at it.

The next morning, I head over to inspect the showers, laundry, swimming pool which all seem to be free. Margaret already has a washing on and instructs me to get our home made washing line out and rigged up somewhere while we have got the chance. Aye Aye Capp'n. I love this camping stuff.

When I go for a shower, I see the biggest insect I have ever seen. I think it is a Praying Mantis and even if it's not I don't like the look of it so, I use the ladies instead. When I come back out, Margaret is talking to a friend, a kookaburra. Her favourite bird is actually eating out of her hand. Her face was a picture. If there was an earthquake at that moment, she would have been oblivious. She would care not a jot. We even got pictures.

MAGGIE:

We stay at Coopernook Hotel car park. It is nothing special but has lots of RV parking. It is great to know we are safe and it is free. We drive on and stop at a pretty wee town called Nabiac and have a wander round. We go into the local butcher. Had a chat and he must have liked us because he gave us two large, marinated lamb steaks and 4 huge rashers of bacon for ten bucks. He also recommended going down the Buckett Way Road and to stop at Barrington just outside Gloucester.

We took his advice and had a lovely drive over to Gloucester. We popped into our local Tourist Information ladies. True to form they were very helpful. They recommended the Billabong garden. I commented that one lady looked very Celtic. She was chuffed and said she recently had her DNA tested. She had Scottish, Irish, Viking blood. She said she might see us at Stroud which has a showground campsite because that is where she lives and she walks her dog there.

We get the Barrington campsite to ourselves and enjoy our lamb steaks. Our timing has been great because all the school holidays are over. It is quieter, cheaper and cooler in February/March.

Another beautiful morning. We have breakfast and sunbathe. It is as if the kookaburras seek me out. One is sitting right next to me on a nearby branch.

We set out for Stroud along the Buckett Way. We see a couple of camels which we heard are very common in the outback. In fact, Australia has the largest population of camels in the world and they are not native.

When we get to Stroud, it's the first time we have to wear our rain jackets and the Aussies are hoping for more rain. Some places we have been to have not seen rain in five years. Pete notices what we think is a possum up a tree. It is grey with a long tail and big eyes.

We treat ourselves to a Chinese and I have a Prawn Thai Curry making the most of the prawns in this area.

Next morning is another beautiful morning. A kookaburra comes right down to see me. Very tame. I am able to feed it from my hand. That has made by day. Stroud is very community orientated and we get free use of the washing machine and the local swimming pool. As we drive on we stop at our pal the bountiful Clarence River and birdwatch pelicans, egrets and herons. We had a brief stop outside Minimi. We had started to walk into the long grass when we stopped in our tracks and turned back. It had dawned on us we only had flip flops on. Not a good idea.

PETE:

It's late when we reach a place called Gosford. We are lucky to find a car park with a forty eight hour parking limit, a surveillance camera and a night watchmen. It is next to the train station and we aim to go to Sydney tomorrow, so we will go by train.

We find out parking is free and park up for the night. It is around an hour and half train journey into Sydney. We are going to visit the famous Bondi Beach. We are on a quest to find the Bondi Vet for my daughter. She had requested that if we end up near Sydney can you get me his autograph? She watches him every morning as she just loves animals. Of course it has nothing to do with the handsome vet.

MAGGIE:

We are near to our target destination Sydney. We decide to take the train rather than drive the busy, city roads. We find the perfect place. Gosford has a commuter car park which allows free parking for forty eight hours. We end up at the top six storeys high. It is very humid and it is open air at the top. There are a few cars parked. It is a bit noisy to begin with, but when we fell asleep we had a great sleep. When we wake up, the car park is full. Ideal for commuters.

The journey was gorgeous and stress free. I love train journeys. My favourite mode of transport. It took about an hour and half to get there. There are five lanes of traffic at Gosford so God knows how many there will be at Sydney. After arriving at Central Station, it is a short train journey to Central Quay. There they were: Sydney Bridge, Sydney Harbour and Sydney Opera House. Iconic.

PETE:

We are heading for the famous Bondi Beach in the hope of seeing the Bondi Vet. As we are sitting on a bus heading that way I notice a sign BONDI VET so we stop the bus and get off. The nice receptionist laughs, "No we are not the famous Bondi Vet Surgery. His was down the road, but he won't be in. He hasn't been there for years. He is in Africa somewhere. What you see on TV is at least five years old." Hmmmmm I need some quick thinking here as I had texted my daughter and told her I tracked down the famous vet. I ask for a Business Card which luckily says Bondi Vet, so I decide to give her that. She'll never know till she reads this book.

We walk a fair distance to the famous Bondi Beach. I am not impressed. It is not that big and it is in front of a city. We don't stay long. We get the bus and then hop on a ferry so we can sail into Sydney Harbour. Our Opal cards got us into Sydney, hopping off and on ferries and buses all for twenty dollars each.

The Sydney Opera House. Now that is impressive. Straight away I think of Billy Connolly standing away at the very top being windswept and interesting. There are all sorts of buskers around it and you are allowed in it. We had a real nice day in Sydney, but it is much, much, much different in the city than the open road. Somehow, we can't wait to get back to the campervan and the joy of not knowing what's coming next.

MAGGIE:

The scale of the bridge is huge. It makes Sydney Opera House seem like a doll's house. It's only when we walk over to the Opera House we realise it is huge too. It is my wish to sail into the sixteen mile long harbour and had purchased an Opal Card each before we left for twenty dollars each. This allows train, bus and ferry journeys to and in the city. How good is that!

We get the ferry to Rose Bay because we had promised Pete's daughter, to try and track down her hero, the Bondi Vet. We are on the bus and Pete spots a Bondi Vet sign. Turns out the Bondi Vet has gone on to bigger things in Africa and now is a bit of a star. We tried, we really did.

We see a sign for Bondi Beach which is not that far away and again iconic. I suppose it was a bit of an anti-climax. I think it is because we have seen so many beautiful beaches such as seven mile beach at Lennox Heads, Byron Bay and of course our personal favourite Burleigh Heads, Bondi Beach is very small and very, very busy. Then I put it in perspective. It is in a city so that has got to be a bit special. A city with its own beach.

We headed back for the ferry. My favourite part of Sydney was sailing back into the harbour near dusk. After walking to Sydney Opera House, it's time to head back as its night time. We must have looked bewildered at the station because a cleaner who worked there came up and asked if we needed help and told us exactly what we needed to do to get back to Gosford. Again someone going the extra mile in his helpful way. Thank you Sydney and thank you kind Aussie man.

PETE:

The next stop we have is totally different. We are driving away from Gosford. Quite a bit up the road there is a signpost saying Australian Wildlife Walkabout. We eventually found it, but it was only half an hour left till it closed. We decide to camp outside for the night. The joys of a campervan.

We are ready to go in when it opens at 9 o'clock. It really was fab. You were allowed to interact with a lot of the animals, as long as you were not noisy. We saw all sorts of animals: kangaroos, wallabies, padymelons, wombats, Tasmanian devils, emus, snakes, dingos, lizards and crocs. We also learned a lot of things such as Tasmanian Devils are a threatened species. They are suffering from a type of face cancer which is caused by them fighting each other over the roadkill which they can smell from as far as nine or ten miles away and which when they find it, they devour ferociously.

There was also a very ancient Aboriginal Site with cave paintings and rock carvings, so we went to visit that. You could feel the ancestors through their art. On the way out, we see a bird aviary, and in one of the cages, we see half a dozen of our old friends hanging upside down, looking quite sad. Fruit Bats. The last time I saw a caged animals eyes like that it was a Silverback Gorilla in a zoo. I

74

felt like cutting a hole in the cage. I could picture their friends flying for miles looking for fruit. I walk away after saying goodbye to them. By that time, the park is closing and to our surprise we had spent all day there. We nearly got locked in. The girl at the entrance thanks us for staying ALLLLLLLLL day. We did learn a lot.

MAGGIE:

We head out of Gosford in the morning. I didn't realise how big it is, as we had arrived in the dark. We decided to just go with the flow. We saw a road named Peats Ridge Road to Calga and I take it. We saw a sign Walkabout: Australian Wildlife Park. We spoke to a lovely Geordie lady at reception. We notice that there are lots and lots of talks and also boomerang throwing lessons all included in the price. Since it is late we decide to come back first thing tomorrow and make the most of the day.

We go down the coast a run and stopped at a butchers. Turns out his ancestors were originally from Paisley and his mum was in the Paisley Buddies. They love the Scots. Got a lovely beef stir fry. Went to the library to use the computers and Pete read an Aboriginal book. Decide to camp outside the Wildlife Park.

We had an absolutely fabulous time in the wildlife park. It was solely Australian wildlife which appealed to us and also there was sacred Aboriginal paintings. When we saw the poster, we realised that, apart from an emu, we had seen every animal on it in the wild and felt very lucky and privileged. Now we were about to get up close and personal.

Firstly, we saw kangaroos, some with wee Joey's in their pouch, wallabies, padymelons and emus all wandering around. Feeding time was due and so we hung around to see that. They came bounding in from all areas of the park. This allowed us to clearly see the wee Joeys as kangaroos are very shy in the wild.

I attend as many 'meet the animals' as I can. The rangers are all friendly and very knowledgeable. First it was the wombat: I knew when I saw it that it could be related to the koala. Sure enough the ranger says in prehistoric times they reckon wombats weighed 300 kgs. Then through the years some adapted to living in trees. Such a cute face on a solid stocky body. As soon as Pete sees the wombat, he knows that's what he saw crossing the road what seems like ages ago. He thought it was a wild boar but it was a wombat.

Dingos up close look quite wolf like. There are two youngsters from different parks for breeding purposes as many dingos have bred with domestic dogs and they are trying to keep the lineage pure here.

Echidnas up close are really like wee porcupines. They eat ants with a very long tongue. They are strange looking wee creatures. We are pleased to learn they are doing well as they are good at adapting.

Got to go and throw a boomerang. We found out it was quite difficult. We also found out what they're real purpose is. They're actually weapons used to bring down prey out of trees.

The first time we see a Tasmanian Devil it looked quite angelic like a small bear. Until it showed its teeth. We found out at the talk that they love roadkill and can sniff it out up to 15 kms away. They make a straight line for it. Two rangers are giving the talk. One has a floor brush to protect from attack. One ranger gave us the nice version (they are her favourites) and the other ranger gave us the Devil version. She said you hear them before you see them, that their ears go bright red and they really do sound and look like the devil if you hear them in the wild. They don't mate for life and they can't wait to offload their offspring.

At the Billy Bandicoot talk, I reflected on the fact I was seeing creatures I had only dreamed of seeing. This was David Attenborough material. Quolls up close. Up close and personal with some fruit bats who were getting blueberries. They are very tame like pets. I felt a bit sorry for them because we know they like to go twenty to thirty km excursions. However, this park is so educational and we have to be thankful for places like this. Lots of animals are endangered in Australia and lots of Aussies see fruit bats as pests. They don't realise they are our friends and we should thank them for spreading seeds for natural rainforest expansion.

We went on the Aboriginal Walkabout to rock and cave art. Could feel the energy. We could feel respect to those who have gone before us and paved the way for our evolution. Paintings and hand prints are 4,000 to 5,000 years old. Feel privileged to live my dream. Still no platypus though.

On the way out, it is closing time. It has been a real memorable day. One I will never forget. All the staff were so lovely. I see a wee platypus keyring and immediately purchase it. The lovely lady at the till takes the time to recommend going to the Hunter Valley wine region on way back up the Hinterlands and "Don't forget to have a bottle of Hunter Valley Savvie Blanc." Cheers Mate!

PETE:

We get our cowboy hats on as we are heading into cowboy country. Destination Tamworth home of the giant metal guitar (same style as the giant prawn) and country music. It's a long drive and we stop quite a lot, so it is getting dark and we haven't arrived yet. We don't want a kangaroo to jump out in front of us so we decide to stop at the first town we come to. Out of the blue we come across a place called Kurri Kurri. It's late and it's dark. We park for the night and fall asleep quickly.

We get up early and it is quite noisy and busy (all Australians seem to get up early.) There is a toilet right in front of us. We are at a public park which is OK in Australia as long as you are not messy or noisy. We make a cuppa, get out the van and Margaret froze on the spot. There, right in front of her, is a giant metal sculpture of…Yep…a Kookaburra.

We didn't even try it. Margaret decides to have her cuppa sitting under it and I take some photos. After prying Margaret away from the poor bird, we take a walk round Kurri Kurri. It was once a mining town and I am an ex-miner so I am interested in the history. They have lots of murals painted on the walls of the buildings and we enjoy walking round looking at them.

After Margaret says cheerio to her kookaburra pal in kookaburra language, we move on to the Hunter Valley which is home to a lot of Wineries. We see fields and fields of grapes and pass loads of Wineries. Then we spot one we know that's the one for us. We sample five and buy a bottle of Sauvignon Blanc as it tastes superb.

The owner has recommended visiting Tyrrel Winery and we do just that. It has a Museum showing the beginnings of the Winery. It started in a tiny hut and developed into the huge place it is today. I go a walk and look over the vast fields of grapes. For some strange reason, I start thinking of my dad. I don't know why. He didn't drink wine and he died seventeen years ago. I guess I still miss him. I make a mental note, "How on earth do they harvest all these grapes, surely it's not by hand?"

MAGGIE:

We are heading up to Tamworth, four hours away. We are going to take our Geordie pal's advice to go via the Hunter Valley. First of all we stop and have Peking Duck in Wyong. We just keep stopping and starting, observing and enjoying. We are beginning to see that the drought has made it hard for farmers, their livestock and also the wildlife. We drive on until dark and then decide to stop at the first place we come to. It's called Kurri Kurri and we know nothing about it. We can't see a thing and park at what looks like a garden area.

When we got up at 7.30 am and I could not believe my eyes. There was a giant Kookaburra outside the van. It was made of metal, the same style as the giant Prawn. I felt as if I was guided there. I read that the Aboriginal spirit totem meaning for the kookaburra is to turn hurts of the past into laughter and happiness. Helping to close and open new doors in our lives. Conquering fears. Take time for healing ourselves and others. I have taken that advice. I feel that every day I see a kookaburra that its guiding me to keeping it real. Listening in the silence to my truth, being honest about my emotions and letting go my hurts which can only be done by acknowledging they are there, deep inside. At the same time reminding me to keep smiling and to enjoy every moment as much as I can.

The town has a slightly run down feel about it. This place has seen better days. There are a few broken windows, a few dodgy looking folk. It kind of reminds me of Ayrshire. Then we get talking to an old guy walking his dog. Turns out it is an old mining area with lots of closed down pits. This guy was an activist and came all the way to Britain in 1984 at the height of the miner's strike. He reckons Arthur Scargill was tame and told us about the 1929 riot and strike in Australia. In fact, there is a mural in town depicting the riot. Turns out there are Murals everywhere. Kurri Kurri has tried to reinvent itself and has lots of murals everywhere. A statue of the miners is going to be replaced by a pit pony and a coal carrying hutch.

We spent a couple of hours walking about town looking at the murals. Then it's time to move on up to the Hunter Valley. I asked Pete, "Do you fancy visiting one of the wineries?"

"Yep" was his instant response.

We come across lots and lots of Vineyards or Wineries as they are known here. We randomly chose to go down Winery Road and come across one we like the look of. It was exactly four weeks ago that I became Mrs Burleigh so we thought it was apt to go into a Winery using a name I call Pete. We got to taste five different wines and if we buy a bottle of one of them then the tasting is free. We try a Champagne, Rose Champagne, Riesling, Shiraz and our favourite a delicious, aromatic Sauvignon Blanc or Savvie Blanc.

The owner is of Scottish descendant and his ancestors came from Orkney in 1850s. He is married to a French descendent and goes every two years to France to buy grapes. He studied wine making for six years at University. He recommends a visit to Tyrrell Winery. He has travelling extensively and told us that the south island of New Zealand is like a cross between Switzerland and Norway. Wowser. Sounds amazing.

PETE:

Still heading for Tamworth but we decide to take the long road there to see what we find. We drive along in our wee campervan enjoying the fact we are living our dream. It's a glorious day—what could be better? We see a poster 'RODEO ALL WELCOME'. It's at Muswellbrook and that's where we decide to head. We find a showground, complete with shows, stalls, running track and a couple of seated arenas. We stop at reception and a woman asks what size of van are you in? That teeny one over there. If you move over to where the big generator is, I'll see old what's his name and if he is in a good mood he might let you camp here, but be quiet.

We have never seen a rodeo before and this is the real deal. We keep quiet and go to sleep with our delicious Savvie Blanc. When we get up, we notice there are a lot more vans and horse boxes. Some of them are real close up to us. We go for a walk. When we come back, our van has the most cowboys I have ever seen in real life or in films like the Alamo. They are leaning against the van talking and yee haaaing with all their horses as well. We really are in the middle of the rodeo. We waited until they had all left before we went back to the van and had a strong coffee. We were in and we couldn't believe our luck.

So off we go and at first it is fun. There are cowgirls and cowboys with lassos. They are lassoing calves. This goes on for a while seeing different horse skills. Then with the music blaring and horses bucking it all seemed good fun, that is until the bulls came on and some of them were mean. They get spurred (jagged) just before the gate opens and they didn't like it. One of the bulls ran all the way to the top of the arena then stopped. All of a sudden, from a standstill, this bull cleared the fence. It had jumped into an empty coral, but everyone was screaming. Meanwhile the commentator on the loud speaker "Yep folks, there's nothing like a live rodeo, there's plenty more where that came from."

MAGGIE:

As we move on to Singleton we are playing a Darkness CD Pete had got in a charity shop. It kind of gave me permission to howl like a wolf. Very therapeutic. We see a sign for the annual Muswellbrook Rodeo. It is a dream come true to see an actual rodeo after being brought up watching western films, especially John Wayne.

Randolph Scott, Allan Ladd, Glen Ford were all heroes of mine as a child but John Wayne was my biggest hero. We are very lucky as we drive into the showground as it is normally designated for horseboxes and people taking part in the show. The guy must like the look of us. He says we can park for a couple of nights. We decide to go a walk round town first and we get wristbands on the way out.

Went for lunch and Pete got to say, "I'll have what he's having," when he liked the look of the chicken schnitzel that the guy in the next table was having. The guy in the bottle shop gave Pete a deal on some ginger beers 4 for 10 dollars. They like us here. Cheers. We had our Savvie Blanc, toasted our marriage and gave thanks for our blessings.

After a great sleep, we woke up excited that it was the Rodeo tonight. We go a walk and come across someone who could have been Roy Rogers with an absolutely beautiful palomino horse.

When we get back, the van was literally surrounded by cowboys, horses and horse boxes. We were right in the middle of the Rodeo. They do love their horses in Australia and are very skilled riders. We didn't like to disturb them. We later saw a girl taking her horse and dachshund dog for a walk together. Pete took a great photo.

The Rodeo starts at around 2 pm. Lots of cowgirls and cowboys, all wearing Stetsons, are lassoing calves. There was an easy going atmosphere at this point and it was all about the horse riding skills and rope skills. This went on for a few hours.

Then it was time for the bull riding and bucking broncos and this is where I started to feel uneasy. I noticed that a belt was put on the animals when they were put into the individual cubicles and this was tightened just before they were let out causing them to be angry and scared and therefore buck. The cowboys were skilful, but I had been a bit naive. It reminded me of being naive as a child loving John Wayne and then growing up and realising the Native American Indians were in fact the good guys, not the cowboys.

As time went on I saw a bull knocking down the door of the stall. These animals are very angry and very powerful. After the rider came off, it went storming up to the top of the ring pawing the ground. I could literally feel it's distress and anger. Then the crowd gasped and a lot of people, me included, stood up. I could not believe what I was seeing. It was a massive, bulky, heaving beast and it cleared the high fence without even taking a runner. Luckily for the crowd it went into the coral where the calves were kept. That could have been horrendous. In fact, it already was horrendous. The final bull came out and knocked the young cowboy onto the ground and then proceeded to trample him. It hit his head. My God I thought he could be dead. He looked as if he was. The crowd were silent. I felt sick. I felt the bull's anger. I have never felt so close to turning vegetarian. It would be a few weeks before I ate another steak.

We decided to stay on the next day and see what else was going on. Log chopping competitions, Huge big Clydesdales. There was a whip cracking competition. We spoke to a couple originally from Aberdeen who moved over in 1988. The love it over here and have never felt the urge to go back home. He worked at an opencast mine and he spoke about snakes. He says most snakes are not really interested in humans, but the brown one is a different story. It will actually chase you and he knows someone got killed by one at his work.

PETE:

When we left Muswellbrook, we passed through a mini Scotland with towns named Aberdeen and Scone. We just had to stop at Uralla as at the roadside there was a statue of a cowboy holding his horse reins while the horse reared up on its hind legs. The cowboy had a General Custer, Buffalo Bill look about him complete with hat. I just had to stand next to him with my kangaroo bush hat on and get a photo. I managed to climb up the statue and stand next to him. His name is Captain Thunderbolt and he is famous for robbing banks and dying in an ambush. Seemingly he was at large for a while. Nothing like a good baddie.

We decide to follow the Thunderbolt route and come to a museum about mining. We have a quick stop and move on. After hearing all that country and western music, it is only fitting that we are heading to Tamworth home of the Giant Metal Guitar.

When we finally reach Tamworth, we know it's Tamworth because of the huge Guitar. I am standing next to it and I am not even up to the bottom string. I get my pic taken next to it and then that's it. We go into the Museum with hundreds, if not thousands, of photos of famous stars. There are no concerts on for a while and we don't have the time to wait. I see Dolly Parton's name and oddly enough, Madonna. There is also Keith Urban and my old pal Slim Dusty of 'the pub with no beer' Margaret is at home here as her attempts at an Australian accent remind me of Dolly Parton a lot.

MAGGIE:

We make our way up the road and come to Aberdeen and to Scone. We end up staying in Uralla, home of Captain Thunderbolt. We treated ourselves to a Sunday Roast dinner of Barramundi and roast potatoes.

Eventually we arrive at Tamworth where there is a huge metal Guitar and a Country and Western Museum. Get our photos taken and then head on. I noticed that Keith Urban has played here and indeed discovered here. In the photo, he is just a gangly, awkward, goofy looking kid but look at him now. Handsome megastar.

Instead of taking the New England Highway we go up the Thunderbolt Highway which is not quite the outback, but quite far up the Hinterlands. We are heading towards Inverell, but we decide to take a detour when we see a sign for Tingha. I remember the lady in the health food shop at Uralla had mentioned that Tingha had a big Aborigine population. We find out that it has only about eight hundred population. Way back in the day, during the gold rush, it had a population of ten thousand people. As we drive towards the village we notice that the surrounding forests are all totally devastated by fire.

PETE:

We were driving along the Thunderbolt Highway on the lookout for wildlife. Hopefully another Wombat. Suddenly, on the road, there is something which looks like a dead snake. Whoa, I said to Margaret my first snake. I need to take a close up picture of it. She stops and says, "you can look out the window. It looks like a brown snake."

"Don't be daft, it's dead," I say. Margaret is saying something like 'hospital' and 'mad'.

"Won't speak again," and I shout back, "it's fine." I am just going to zoom in. She is pointing and shouting and as I turn round the snake flips upside down and in the air and coils itself, not once but twice then starts moving. "Fkn Idiot," someone says next to me.

When we are talking later to a guy at the campsite, he says, "Mate, I stay here and I wouldn't go near a brown snake, they are known to run after you, even get in cars." Margaret looks at me, then asks are there any brown snakes here? Nope. All the black ones eat them!.

MAGGIE:

The campsite at Tingha is one of the nicest and cheapest sites we have been on. It is community run. It is not really a tourist area but it is famous for fossicking. Some people are still finding sapphires up at Inverell. Quite recently a guy found thousands of pounds worth of Sapphires.

The manager, is really friendly and helpful. We had mentioned that earlier on we had left our phone charger in the Mining Museum. Later on that night he knocked on our van with a phone charger for the cigarette holder. We told him about Pete's snake encounter I said it is a good job it was badly injured. I had remembered the snake story from the couple the week before. It did look dead but it was alive. He said there is no way he would have gone out to look.

When we go for a drive, we go down three different roads in three different directions and we see the same devastation. No trees left. Just charred stumps. No wildlife. Just an eerie silence. The smell of burned ash is still hanging in the air and the fire took place four weeks ago. I feel as if we are seeing the real Australia up here away from the coast where drought, bush fires, flash floods frequently cause havoc.

Pete decided to go into a shop to see if he can buy a plug in adapter as we left that as well. When we describe what we are looking for the young guy says, I think I have one from years ago through the back. Yep you can have it. Save you having to buy one. Thank you kind Aussie.

We then spoke about the devastation we had seen. He showed us footage on his phone. He took it the day before the fire arrived and it was scary stuff. We could see big smoke clouds making their way towards the village. He said on the day it arrived it felt like someone was blowing heat and smoke onto his face with a hairdryer. The heat singed his hair. He had never seen one this bad, this close up. Place was like a tinderbox. He told us people are not taking the same precautions as his dad did twenty years ago. He took all the trees away from around the house and created a fire break. He said the Aborigines used to safely clear the tinder grass with small contained fires. Now they have outlawed this practice, instead they have years of growth which goes up in flames so fast with huge consequences killing and destroying everything in its path.

PETE:

On the drive up to Tingha, we notice nothing but burnt trees on either side of us. In fact all around us. We had heard about the forest fires and had seen them on TV back home, but to see them first hand is quite devastating. A local guy showed us his mobile phone with a film of it burning round the village. He said he could feel the heat on his face like a hairdryer blowing. The woods on the north, east, west and south were all burnt. It was a miracle that the village was saved. He told us we should go a walk through the woods and see what we think.

The fire was weeks ago but we could still smell the charred wood. It was very, very quiet. You could actually feel the pain of the place in the air. Very sad. As I turned to go back to the campervan, I could see a wee shoot of green reaching out through the charcoal. New growth.

The little campsite was great and folk were very nice. The area is well known for fossicking. There were old tin mines in this area. Fossicking is panning for crystals. We are invited to go fossicking with Travis in his pickup. He arrives armed with a long handled shovel and a riddel. He looks like a nice chap, but I can't stop thinking about those films, where you get whacked on the back of the head with a shovel and disappear. Thankfully, he jumps out and begins to show us how to pan for crystals. It's fun and we even found get a couple to take home with us. We thanked him and said goodbye. Love this wee site. It had mats to stand on in shower, soap for hands, tiled floors. I felt pampered. Living a simple life lets you appreciate things that most folk take for granted.

We head for Inverell and decide to keep on driving. It is real cowboy country. Looks like a prairie and we can see that there has not been much rain in these parts.

We are heading to Glen Innes and just before we get there we come across an old Aboriginal site. It is quite dark. We stop the van and walk over to the site. We notice there is a clay pigeon shooting area. It looks like it's been closed for a while, maybe the clay pigeons have migrated. As we walk over we actually sense the area we are in, rather than see it. I hear a humming noise. There are no electricity wires and we are standing on rocks. There is an odd feeling here. We don't like it, so we move on. I make a note to find out what happened on this site.

As we drive along we come to a camping lay-by which is next to a Stonehenge lookalike, with two giant standing stones and one across the top. We get a photo of the sun setting behind it.

Next morning we head into Glen Innes. It is quite a big place and we drive up to a hill. We come across an exact replica of the Orkney Standing Stones. It has latitude and equinox times. Everything is written down on stone to read. We then realise that without even trying we have landed here on the actual Autumn solstice here in Australia (Spring back home). We decide to camp there for the night so we can get up early to bring in the dawn and celebrate the Solstice.

Just before sunset we decide to pick a card from the Native Spirit cards. I drew one which had a picture of Standing stones on it and if that isn't enough Margaret said lets open a bottle of ale to celebrate. We happen to have it in the van for a couple of weeks. When we get it out, the label has a picture of Celtic standing stones on it. You could not make this up. All we needed was Merlin to walk out from between the stones.

Maggie:

After we have stopped at a strange Aboriginal site where it feels like something bad had happened, we swiftly move on. We come across an ideal camping lay-by with a Stonehenge next to it. I immediately notice a Poster saying it's Harmony Day tomorrow with lots of different cultural dancing and food. I really would love to see a genuine Aborigine Dance. I notice it starts at 10 am till 1 pm. I also notice there is a huge full moon.

We went up to the Standing Stones for breakfast. We asked at the Croft Cafe about the story of the standing stones. They were built in the 1990s and they are very proud that they are a replica of the famous Orkney Standing Stones.

Ended up missing the Aborigine Dance by ten minutes. I tried not to blame myself for missing it. This is a habit of a lifetime me feeling the need to blame, usually myself. There is nothing I can do to change what has happened. I get over my disappointment, but I feel time is running out to see authentic dance and hear someone play a didgeridoo. Instead I see a Nepalese Bollywood style dance and get to taste various international dishes. Very nice it is too.

We decide to camp for the night at the Standing Stones so we can get up for the equinox. We had not planned to land here on 21st March but we are going to make the most of it. The moon energy is very strong and we saw big orange moon rise over the standing stones. To celebrate I had brought out an ale which just happened to have Celtic standing stones on its label. We set the alarm to get up and see the sunrise. The Celtic stones feel authentic and it is obvious it is a labour of love. A lot of thought has gone into it. It is lovely to be up early enjoying the peace, the views and the wildlife.

PETE:

As we walk around Glen Innes and get speaking to a few of the locals, it is obvious they are pretty proud to have Scottish blood in them. Some of them know their Scottish history better than we do. They know the convict ship their great-great-grandfather came over on and regularly have Highland Games etc. In fact one old guy says, "You're in luck. Today is Friday and every Friday at 12 noon a man comes out dressed in traditional Scottish kilt, the full outfit and plays the pipes. It is a quarter to twelve just now he says you better go if you want to see them."

We head down eagerly. We wait a few minutes and as nothing is happening I decide to ask the next passer-by. A big guy with a hat comes towards us. You can tell a mile away he is Australian. "Excuse me," I say. "Is there a piper about to play the bagpipes?"

"You what, mate?" he says, "Pipes! I hate the bastards! Let me tell you, Son," his eyes narrowing. "The first time I saw them was on my daddy's kitchen floor. 'What the heck's that, Dad?' I asked. 'I dunno, Son, but stand back'. I gave it a poke with a stick and it yelped, so I reckon it's some kind of wild pig." Then he walked off laughing. I think we made his day. He probably waited years to be able to say that off the cuff to a real Scotsman. It certainly made our day.

MAGGIE:

We went to get some strawberries and the guy let us pick a small bowl for free as it was near the end of the season.

We spoke to a guy who was part Aborigine and we had an interesting talk with him. We told him about the eerie feeling at the site we had visited and we reckon that something nasty has happened there. He says he had a similar experience whereby he felt three taps on his shoulder. He thought it was his wife, but he turned round and no one was there and he knew it was time to get out. Gotta follow your instincts.

He told us that his grandmother was Aborigine and when she was widowed she was allowed to stay in her house she had bought because she had married a white man. However, all her siblings' children were all taken away from them for a 'social experiment'. They still live in a mission area where drugs, alcohol and poor self-esteem abound. He was lucky because he had one white parent he got to go to university and has had a successful life through self-belief, work ethic and talent. Sadly his relations have no such self-belief.

Just by being Scottish, we feel like local celebrities. A lady who found out we were from Ayrshire asked if I knew such n such from Ayrshire as her relation from Gretna plays curling with him. After we saw the piper play at noon, we got talking to a guy and he told us he got ceremonial garments passed down from his ancestors but shhhhhh it might upset some people. Reckon it was the Orange Lodge. He says he is the grandson and he has never wore it, but he has kept it.

I go to the library to get some quotes for our new hire campervan in New Zealand. I asked why the Aboriginal Centre had been closed down. The librarian said they had been no profit made. As it is a huge part of Australian history, I feel that this should be like Museums a non-profit making exercise. If funded by the government surely, it could be used to try to help bring back self-esteem to the indigenous people? We certainly would have been very interested to visit such a Centre.

We went back up to see the Standing Stones one last time before we leave. We met a couple who had just been up to Lightning Ridge way up in the Outback. They said it was heart-breaking. Lots of towns are boarded up and deserted due to the drought. It is just too difficult to make a living there. The spoke about how difficult it has been for the farmers because of the drought and also the fact up north the opposite happened and 500,000 cows were dead because of the flooding. Many pedigree cows lost and many farmers committing suicide. We had been in a non-news bubble I said. They told us a huge cyclone was hanging over the north west and towns were being evacuated.

The weather is the boss and the past week has been quite sobering and a reality check as to how difficult life can be here. However, there is a huge community spirit and kindness in Australia. Most folk try to help each other the best they can. Certainly they are not whingers or whiners.

We get talking to another couple who love our Scottish accents. We remind her of her dad she tells us. Her husband says, "I couldn't understand him either." Another family come up to speak and the son has white Celtic alabaster skin. He says he prefers the cool air of Scotland. I had noticed that not many people have a tan in Australia. I think it is because they live in dread of skin cancer.

PETE:

Well! After leaving our friend the Aussie bagpiper and saying goodbye to the Standing Stones we are driving along heading to Tenterfield where we go into the Information Centre. I happen to mention we are hoping to see a platypus. The man winks and says, "Do you want to see one? Follow me but you gotta be quiet." We silently follow him then he stops and points. There you go-one stuffed platypus! Sure enough there is a stuffed platypus on the table. "Go ahead. It won't bite." It did look odd with its big duck bill and webbed feet. It would be great to see a real live one. That is our wish to somehow see a platypus in Australia and a kiwi in New Zealand. Don't know how we are going to manage as both are very rare. Most Australians have never seen a platypus. We decide to do our trick and 'put it out there'.

MAGGIE:

We are on our way to Tenterfield and then Casino as recommended at the Winery in the Hunter Valley. We stop at the Information Centre and they have a stuffed platypus. Looks like that is as close as I will get to seeing one as time is running out. I touch the platypus. Its fur is quite soft and its tail is rough. Seemingly a male platypus can give a nasty poisonous wound with its hind foot spur if you go too near.

The kindly gentleman in the Centre warns us there has been more fires on the road we are going. He says he was in the UK for six years in his twenties and shhhhhhhhh but there is not a day goes by that he does not regret coming back to Australia. I think the extreme weather gets to him. He is still good natured and kindly as we have come to expect in this wonderful country. He tells us he goes back frequently to the UK with his wife.

We decide to press on as we don't have much time to stay and explore. We see the devastation of the fires once again. After driving quite a distance, we pull into an overnight camping area just fifty kilometres from Casino. This proves to be very timely as ten minutes later we can see there is a huge storm brewing and sure enough it lasts for seven or eight hours. Lightning, thunder and torrential rain. Hallelujah! We are so happy for the wildlife, livestock and people of this area. Everything seems so refreshed in the morning and everyone has great big smiles on their faces.

PETE:

As we drive on we encounter our second electrical heading our way. We manage to pull into a free camping lay-by with a dunny. For those who don't know a dunny is—it's an open air toilet. Basically it is a wooden shed with a door and a seat on top of a hole in the ground, a big hole. They are pretty common in Australia and believe me, in some cases it is a good job they are there. Surprisingly they are not as bad as you think, although you only go when you really need to, but it is freedom camping.

Within minutes, we can't see a thing for the heavy rain. As the Electrical Storm gets real close once again we get really mesmerised by the light show. After the last one, we realise that it is highly unlikely that the van will be hit by lightning. We watch the forks of lightning light up everything for miles around us. I suppose that's why it is called lightning. It is still quite scary. We do manage to get some sleep. In the morning, a lady nearby shouts over, "Wasn't it great? Some people haven't had any rain for seven years!"

MAGGIE:

As we drive towards Casino we see much more green grass. A joy to behold. We find out that Casino is called the cattle capital of Australia. I think it is because it has a massive Abattoir. We stop at the Information Centre and get speaking to a lovely, helpful lady who explains that three weeks before there was no grass. The Abattoir was full every day. The farmers were forced to get their cattle slaughtered because there is no food for them. It suddenly dawns on me that is why meat is so cheap here. The market has been flooded with it. Then the rains came. Mother Earth has recovered really quickly in three weeks. She never fails to amaze me.

PETE:

Off we go and jump into our good old faithful George the campervan. I must say it is great travelling around Australia in a brightly coloured campervan with a name and which actually has Hippie written on it as well. I feel so lucky. We have not had any television on or DVDs on in the van. Well one Nelson Mandela DVD but that's it. Well, we love it. We are heading back to Burleigh Heads before we say cheerio to our van. First onto a place called Casino and then Mullumbimby. Go for it George. TOOT TOOT, HONK HONK.

Well, nice one George. We are parked outside the Info Shop at Casino and reliable George decides to have a flat battery. To be honest I don't think he wants us to leave as we have had such an amazing journey together…but all things must pass.

MAGGIE:

We meet and speak to a lovely lady and hear how, as a Philippine woman, when she arrived, she found it very difficult to be accepted and to get a job. She had left a good job in the Philippines because she fell in love with an Australian farmer. Her qualifications were not recognised in Australia and she had to start from scratch. She also had to get married at the beach in Australia as a Philippine wedding is not recognised in Australia. She happened to say her name is Shona. We say we didn't expect her to have the name Shona. She says she has no idea why her parents chose a Scottish name. She gives us her card and invites us to come and stay a couple of days. She is such a lovely lady. We wish we had the time. Hopefully one day we will be back.

PETE:

We wave down a policeman and he says, "Sorry, mate, can't help." I think they are all busy, there's a good few of them. It might be the Election. Margaret has been speaking to a kind lady and she helps us out by phoning her husband who then phones his local garage and who sends a young lad to help.

It's not long before we say many thanks and G'day to Shona and to the young mechanic, start George up and away we go to Mullumbimby.

MAGGIE:

Keep on trucking to Mullumbimby. We stay at a lovely campsite overlooking the mountain. It's lovely to see the sea again and can smell it too. Get talking to a Geordie woman, a Sunderland supporter and end up talking about when Sunderland won the cup in 1973.

We see our bat friends. We are so happy to see them, a chance to say farewell. We realise that we have only seen them when there has been plenty of fresh water enabling fresh fruit to grow. I also saw and heard a kookaburra. I'm gonna miss them. They are forever in my heart and memories.

PETE:

Three things happened to us in Mullumbimby. It is our last stop before our return to our favourite Burleigh Heads.

Firstly the return of the fruit bats. Man those guys will stay with me forever. Again, there were thousands of them flying not so high above us. So good to see them again as we will not see them in New Zealand. To Margaret's delight there were a few kookaburras and we won't see them in New Zealand either. This must have been a great surprise for Margaret seeing the bats and hearing the kookaburra seemed like laughing bats.

Secondly I got to do a real Bush Tucker thing. We met a woman walking up and down among the trees in the local park. I asked if she had lost something. No she says I am collecting dinner. I told her I was a herbalist and can I ask her what she is collecting? Turns out she is collecting Bunya Nuts from the Bunya Tree. Follow me she says and I will show you what to look for. The first thing she says is to watch your head as these nuts are inside large cones. If one hits your head, you will see stars all night. She showed me the cones, buried beneath masses of fallen leaves the same colour as the cones. So that's why the stick in her hand came in handy. Not only does it brush the leaves aside, but it protects you from spiders. The last thing you want is to be bitten by one of the many poisonous spiders in Australia. I will be glad to get to New Zealand in that sense. At least, there are no poisonous bitey things over there. God, it is bad enough suffering from insect bites here. My oil works slightly so I am hoping New Zealand is kinder to me that way.

As I collect my Bunya Nuts the woman tells me that I've got enough to make a small meal. You crack open the cones and inside are the nuts. They are about the size of a chestnut and just as hard. She tells me that you have to boil them for about twenty minutes, then fry them in butter and voila they are ready. She then gave me her collection to take away in a bag for Ron. LaterRon that is. Very kind.

The third thing was a real scare. It was not until we were miles up the road that Margaret realised she had lost her gran's locket. Oh no it could be anywhere as she had a shower at the campsite as well as there being grass everywhere. We searched the van under, over and inside everything. No locket. We had to drive back to the campsite. We asked the owner if anyone had handed it in and when the answer was no she retraced her steps. Meanwhile, I decided to do my Sherlock Holmes and went to the spot where we had camped. I recognised a few wooden posts because of their shape and saw the patch on the grass where the van had been. So I pretended I was putting out the chairs where we sat. I sat in the chair where I thought Margaret had sat and there it was. Like finding a needle in a haystack. I surmised that she had sat the locket in the chair for safe keeping while at breakfast and somehow just folded the chair without thinking. Anyway all's well that ends well.

MAGGIE:

We have headed to Pottsville because we like the name. I realise when I get there that I have lost my granny's locket. We realise that it must be at Mullumbimby campsite and so we turn and go all the way back. I am slowly learning not to blame myself because it does not change anything except make you feel like crap. It is so humid I am sweating buckets. It's like looking for a needle in a haystack, but Pete is confident he will find it in the grass and he does find it! I am so grateful Pete thanks.

We have a T-bone steak Sunday dinner and head back to Pottsville. We go to the beautiful beach and I let the wind blow through me. Sitting in the car a Ranger came up to advise us that we could not stay the night. It ended up, yep, he loved Scotland.

The various campsites along the road are too busy or too expensive so we kept going on to Burleigh Heads. It felt like coming home.

PETE:

Well, it's been a real journey and here we are back at Burleigh Heads. As we are married now we don't book the luxury suite. We camp in the very big and popular campsite and decide to walk everywhere.

We have two full days left then we are off to Brisbane to hand over ol' faithful George on the fortieth day of our trip. Man, I just wish we could have seen a real platypus or some real didgeridoo playing Aborigine. Now that would have been the icing in the cake.

I get to cook my bush tucker on the outside stove at the campsite. I boil them, then as the shell splits I take them out and fry them. They taste quite piney and since I am wearing my bush hat I feel right Bush Tuckery.

MAGGIE

There is a campsite nearby to Koala Park. We realise that Burleigh Heads is a very popular holiday destination for Australians. The campsite is huge, but it feels so good to be here. Next stop is Koala Park to see Dianella our hairdresser and also to do the laundry. There is another campervan here and I am not too sure about the guy. I think he may have been alone too long or has mental health issues.

Next morning we are outside in our campervan when a lady makes a beeline for us. Turns out she has a campervan, I think she had also been too long by herself and she was quite hyper. Reckon she is also a poor soul in need of help. There are two sides to the camping. There are the grey nomads who out of choice give up their house to travel and experience the great country they live in and then there are some folk who live out of necessity in their van. We feel lucky that we had the choice to be grey nomads.

I have spent my time in Australia really hoping to spot a platypus in the wild and to see a genuine Aboriginal dance and/or a real didgeridoo being played. We fly out to new Zealand on Saturday so it is looking unlikely.

PETE:

We were walking into town we past the Aboriginal Culture Centre. We might as well go in says Margaret. We have been in a couple of times before and they recognise us. When we are in, to our delight, we find out that there is an Aboriginal Educational Class on tomorrow and because we have shown so much interest we are told if we turn up tomorrow at 9am we can sit in with the class. That is a God send as it is our last day at Burleigh Heads tomorrow. On the way out of the building, I notice a poster with Aboriginal people covered in their traditional body paint and playing a didgeridoo. I really hope it's them. We are excited all night and get up at the crack of dawn.

When we arrive to our dismay, there is no sign of any class. We get told that due to unforeseen circumstances it has been cancelled—it was too good to be true I say to myself—then Margaret comes out with a huge smile. We have to come back at noon for the display, by this time she is laughing like a kookaburra. You couldn't make this up. It doesn't get any better. Well actually it does...

MAGGIE

We popped into the Jellurgal Aboriginal Culture Centre which is across from Koala Park. This Centre is specifically used for educational purposes and it is part of the school curriculum to learn about Aboriginal history. It feels progressive and they realise we are really keen to learn as well. I cannot believe our luck the guy at reception says if we turn up at 9 am tomorrow we will be allowed to sit in with a school visit to hear the didgeridoo being played and a see few traditional dances performed. I asked him about the David Fleah Wildlife Park and he says that they have an actual platypus and that it is very rare to see them in captivity never mind the wild. I think my wishes are coming true tomorrow.

I did a Native Spirit card this morning and got Eagle Medicine with a beautiful picture of an eagle soaring above mountains. When we arrived at 9 am, the guy apologised that due to unforeseen circumstances could we come back at 12.15 pm as it has been postponed till then? Certainly! We are just delighted to have this wonderful opportunity. He advises us that we will be able to go in to the David Fleah Wildlife Centre come back out for our visit and then go back in again to the see more wildlife. He offers to phone them up to explain what is happening and just to mention his name. Thank you to the kind staff at the Centre you were so helpful and really made our day a day never to be forgotten.

PETE:

To pass the time till 12 o'clock we went to the David Fleah Wildlife Park. He was a very famous and well respected here in Australia for his conservation work. Guess what we get to see? Only one of the very few live platypuses in captivity. We are up there quicker than a shooting star.

We get to see a wee platypus called Wally. They have built a nocturnal type of habitat aquarium which fools Wally into thinking it is night time. We can see him come out frequently to hunt. Cute as he is, you don't want to get on the wrong side of him as he has toxic spurs defence that he squirts at predators. So toxic it can kill a dog. We see a huge crocodile and loads of kangaroos, koalas then it's time to go down to Jellurgal...

MAGGIE:

I saw Wally the Duck Billed Platypus in a nocturnal setting as near to his natural habitat as possible. He is adorable. One of the most unusual animals on the planet with his duck bill, web feet, furry body and flat tail. He was just a baby when rescued in a car park and is half the size of a normal male. Wally is safer at the park as it is doubtful he would survive in the wild. It is Australian policy that wildlife should be living wild when possible. I could watch him for hours and I make ooh and aah noises regularly. I feel quite emotional. He thinks it is night time and seems really happy darting around his aquarium.

A young Japanese family come in and when they eventually see Wally they also ooh and aah in Japanese.

The Birds of Prey and birds of Queensland show was amazing and an eagle flew so low I felt the breeze on my neck from its wings.

PETE:

On stage two, traditionally painted Aborigines come on, one of them looks quite young. The other guy is the one from the poster I saw earlier. The young guy does the dance of the eagle. We learn that real didgeridoos take years to make as termites eat the wood to make it hollow.

He explains the ceremony he is going to do first. The white stripes across his painted belly represent the coming of the storm clouds and the sound of the didgeridoo is warning the animals what is coming. He goes on to play a mesmerising low humming tune without once stopping for a breath. To Margaret's absolute delight he starts imitating the kookaburra on the didgeridoo. Man that bird has been following her all over Australia. When he finishes he asks, "Would anyone like a go?" and I am up there in an instant but I don't even manage to get a peep out of it.

We meet them outside and get talking. They like us. We get our photos taken and say cheerio and good luck. To my delight Margaret has videoed the kookaburra storm ceremony. It is our last day in Burleigh Heads and we get to see this authentic ceremony.

MAGGIE:

We leave for the Jellurgal Aboriginal Centre and get to see traditional dance, the first being an Eagle Dance. The guy explained that Eagles are very important to them as they believe that the Eagle leads from above and takes them to fish. We also get to hear an amazing didgeridoo ceremony.

Afterwards Pete gets onto the stage for a go, but his is silent. We agree that the Scots and the Aborigines have things in common in that we have been losing our culture and language and now it is in our school curriculum hopefully being preserved. We get to speak to one of the Aboriginal guys outside the centre and I say, "This is your time."

"I hope so," he says. Since introduced to sugar diabetes is rife among the indigenous people as is alcohol and drug abuse. Sounds so familiar to Scottish problems back home. As if we lost our identity, we gave up hope and became complacent. I still play the video of the didgeridoo regularly to this day. I find it very uplifting and healing.

We have said cheerio to our lovely Dianella when we nipped over to her salon. We are now going to head down to the 'rump and a pint pub' to meet our lovely Celebrant. She has our Marriage Certificate and we are going to have a catch up. On the way down, a young guy walks past and randomly points up to a lamp post and says, "G'day there's a sea eagle sitting up there." Sure enough we look up and see the eagle. That is three eagle encounters I have had today. Seems to be strong Eagle medicine going on.

PETE:

Away we go with our hearts full of joy and we go over to see Dianella our wedding day hairstylist and then to see our Wedding Celebrant. We tell her we are going to come down and pack and clean George in the morning at the beach near little Burleigh Hill. At the place we got married.

We woke up early and head down to our favourite spot, for our last glimpse of Burleigh Heads. We decide to make breakfast and as we bring everything out and are cleaning the van, a security man comes up, "What ya up to, you can't stay here overnight." We explain our circumstances and tell him we are heading to Brisbane then New Zealand soon. He breaks into a huge smile, "That's great, you'll love it. That's where I'm from. In fact, I'm going to give you your first Māori lesson" *Hikatta, Heekatta, Katta, Kattah*…or something like that. We recognise it straight away. It's the same thing our friend said to us: the little woman who appeared from no-where and put her hands on our heads. It seems she is not so daft after all, she was, in fact, giving us a Māori Blessing. The security guy said it means BE STRONG, SAFE JOURNEY and other nice things to protect us. I like her although I am still a bit dubious about her saying she stayed in Bethlehem and Judea. Well, she did get bitten by a snake.

MAGGIE

We get up at 5.30 am so we can go down to pack and tidy George out at the beach near Little Burleigh. Then we can go to our wedding spot, go for a walk on the beach and head to Brisbane. Today is 27/3/2019—the 40th day in our campervan George and the day we hand George back. As we are tidying up a security guard comes over as it is illegal to overnight park at the beach. When he heard our story and that we are heading to New Zealand, he begins to wax lyrical. The North island is where he is from and he says that he would go back in a minute if it weren't for his boys living in Australia. He said be sure to make our own hot spring bath at the beach near Rotorua. As he left he shouted out a Māori saying. Me and Pete both look at each other as we realise immediately this is part of what our wee woman said to us when she put her hands on our heads all those weeks ago, the day before our marriage, in the Sacred Rainforest. He says it means STAY STRONG SAFE TRAVELS. We find out that she has given us a Karakia which is a Māori blessing, incantation and prayer used to invoke spiritual guidance and protection. How lucky are we? We really do feel very blessed.

PETE:

Away we go and hand over the keys for George. Then back to the same Airbnb in Brisbane. Since we stay for three nights we get to know the area quite well. We find a place where the coffees are amazing.

We see the same Aboriginal guy, sitting in the same place with a guitar every day. He has never played it any time when we see him and he and his friend seem to go into Dreamtime. It is as if there is another level as they tilt their heads and calmly stare into space. Not even blinking. As we pass on our third day Margaret says I am going to ask that guy if he actually plays and sure enough she does just that. He grabs the guitar and belts out Elvis looking Margaret straight in the eye and both of them laughing.

We also head to the Brisbane river and find out that there is a free ferry service which takes you up and down the river all day passing under the bridges and taking in the skyline. We jump off and come across a replica of Leonardo Da Vinci's model of the first helicopter he invented. It is a working model and we need to push hard to make the blades move. We then come across a field of giant astronauts in spacesuits which is something to do with a space programme here. I happen to notice on the way back this area is called Kangaroo Point.

MAGGIE:

Our forty days are up and our fortieth night is back where we started at Brisbane in the same room.

We say our fond farewell to George our Hippie Van. When we get to the west end of Brisbane, we find out that not only has our hostess broken her foot, but she has also opened her Vegan Tapas Bar.

The choice of restaurants in the west end of Brisbane is great. We decide to go up and try the Vegan Tapas. We had Roast Carrot and red pepper hummus with a pomegranate glaze and roasted hazelnuts followed by a chocolate orange tart with orange slice dipped in chocolate. Every bite was zinging with flavour. My taste buds were very happy. I embarrassed Pete by mentioning he liked steak. I was out of order. Not thinking. The Sol Bakery has my vote for my favourite coffee in Australia. A cappuccino loaded with real chocolate on top.

We had a Mexican meal on our last evening. The two waitresses were so adorable. They look so young. One girl is studying sports psychology and hopes to do her masters. They take great pride in the authentic food and are delighted that we are enjoying it. Corn cupcakes, fried tacos and pork achiote. We ask about the skeletons on the wall and they explained about the Dia De Muertos (Day of the Dead Festival.) These sweet girls have really inspired me to hopefully, one day, visit Mexico and see it for myself.

The Australian folk have been so friendly. One Aborigine guy had been sitting at the same spot, with his guitar for the past three days. I had never heard him play and on the third day he asked for money. I asked him can you play your guitar? He starts up and belts out Elvis Make The World Go Away. We both looked at each other, his face lights up in a big grin and we both laughed. I handed over some money. Mutual respect. Priceless.

We go on the free boat ferry service and hop on and off. Brisbane is a very laid back city compared to Sydney. It has a magnificent skyline. The view from Kangaroo point is amazing. It was Pete who noticed we were at Kangaroo Point. We have gone full circle I say. I explained that the student girl sitting next to me on the way over on the plane had recommended this very place to visit. We had ended up there without even trying.

We buy a bottle of Prosecco to toast our time in Australia. Again the skyline view from the flat is amazing. We have seen so many weird and wonderful things in Australia. It's hard to choose my favourites. We were in a paradise bubble for five weeks then we went up into the hinterlands and saw the devastating effect of drought and forest fires. We also saw how quick mother earth can regenerate when given the chance. Seeing eighty year old turtles on the beach and little ones running for their lives was very humbling. The beaches, the people: so kind, so friendly and funny, the amazing birds and animals, getting to see a live platypus so cute it brought tears to my eyes and on the same day getting to see the Eagle Dance, hear the didgeridoo. Finally, getting married at Little Burleigh Beach to such a lovely guy as Mr Burleigh, These memories will be with me, in my heart, forever.

PETE:

Our forty days and forty nights are over and so is our Australian trip. I say cheerio to insect bites and a real weird and wonderful adventure. New Zealand here we come. Wonder if I'll see any Lord of the Rings places I think to myself as we head to the airport. I finally look back. It's been awesome.

Our new journey has begun. I thought it only took about an hour in the plane but New Zealand is actually three and a half hours away. I could fly from Scotland to Malta in that time. After a while, as I look out the plane window, I see stunning scenery below. Lush mountains and hills, turquoise coloured sea. I can't wait.

MAGGIE:

That's us in Christchurch. The flight took three and a half hours plus three hours' time difference. We left at 7 am and arrive around 1.30pm. We get picked up and taken for our Kiwi campervan. Actually it's a Tui and his name is Noah. He is a dinky Toyota automatic and feels good to drive. I immediately sense that the people here are a bit more reserved than the Australians.

Our first night we find a campsite for five dollars each. It is run by the government to encourage sensible camping. Toilet is clean. No shower. Fair enough. We woke up to a beautiful Autumnal sunny day. Not a cloud in sight, but around ten to twelve degrees cooler than Australia. We decide to head to Kaikoura. The drive from our campsite starts out feeling and looking a bit like Scotland.

PETE:

In the Airport, I was surprised to get the soles of my shoes checked. I soon discovered that this happens wherever you go. The land is so delicate and precious and so many species of plants and animals have been brought in from other countries they are actually killing off native plants and birds. As I walk on, a black dog comes up and sniffs. I think it likes me then I realise it is a sniffer dog. It seems to hang round me for an eternity and I get the jitters. Please, please, please I hope nobody sneaked something into my luggage. However, all is well and the handler smiles and says kia ora. There is a definite different vibe here than in Australia. I like it though. I can't help wondering if the insect bites will stop.

We get our new campervan. He's not a Hippie, but he's called Noah. As I have a grandson with that name I am well pleased. We reach our first camping area. It is very spacious and as we are tired because of the travelling and three hour time difference, we have an early night.

MAGGIE:

On the way to Kaikoura, we pass a sign for Auchinleck. I do an emergency stop as I was brought up on a farm near Auchinleck in Scotland. As Pete is taking my photograph a woman comes to speak to us. She is intrigued as to why we want a photograph of the sign and we are equally intrigued to find out why it has this name. Turns out is it's after a guy called Francis Auchinleck (pronounced Auchinleek). She also tells us that it was not just Christchurch, but this whole coastal area was badly affected by the devastating earthquake in 2016. It has taken up till now to rebuild the railway and repair roads. As we approach Kaikoura there are still roadworks. Now we know why.

PETE:

Noah, our van, is small and comfy. We get some sturdy banana boxes. They are very strong and make great storage drawers for underneath the seats or bed. It saves space as well. In the van, we look at the atlas we were given when we got Noah. We notice there are wee symbols which indicate Lord of the Rings locations. They are zigzagged all over the south and north islands. We decide to try and see as many of these locations as possible.

We also found out there are lots of designated freedom camping grounds. They have honesty boxes with envelopes provided and the fees are very reasonable and sometimes free if the only facility is a toilet or a dunny. Both New Zealand and Australia welcome camping and try to make it affordable. They realise that campers spend their money in local shops and bars therefore helping the local and national economy. I wish Britain would wise up to this as well and help make camping so affordable and welcoming.

We're driving along and "Oh No" I feel an itch on my leg again. At least, there are no poisonous creatures in New Zealand, but I have found out there are plenty of sandflies and their cousins. I put on my home made oil which helps a little. I am definitely going to ask locals what they use.

As we drive along we pass a sign that makes me look twice. AUCHINLECK. This is the name of the local village back home on the other side of the Earth. We have got to stop for a photo.

As we are taking photos, a woman walks over. She explains the name is actually a person's name. She says when the earthquake happened three years ago it closed the whole coastal road and railway. This took a long time to rebuild and supplies had to come by helicopter. We realise, once again, just how lucky we are in Britain.

MAGGIE

As we drive the mountains are getting higher. The guy from the Hunter Valley Winery in Australia was right: Kaikoura is stunning like Switzerland meets Norway meets Iceland. The water is a pure turquoise from the glaciers.

We go for a walk and then for a blue cod fish supper. Freshly caught and delicious. Thought the girl serving us was Scottish, but it turns out she is Swedish, but lived in Scotland and learned her English there. She sounds cute.

Pete had spotted a council campsite a few kilometres from Kaikoura as the local one was very busy. Night is falling and the stars are out en masse. Felt as if I could touch them, reach up and hitch a ride along the Milky Way. The mountains feel as if they are meeting the stars, so clear and so near.

PETE:

We say cheerio and hope to find a spot to camp nearby. The friendly lady told us, if we are lucky, we would see whales. She told us that she has actually, frequently, seen pods of them. We drive on and I see a place literally just off the road that looks suitable to park for the night. I check to make sure there is a toilet. Then head to Kaikoura just a few kilometres away to investigate.

As night falls we end up back down the road to the little campsite. Journeys tend to be long here, so if you have been driving for a while it's wise to stop when you find a suitable place.

I've noticed the trucks and lorries in New Zealand show no mercy. They're right up your arse and toot and honk impatiently. That's another good reason to stop.

We pick a spot in the dark that we think is fine as we can't see much. Later that night though, the stars come out in their millions. I've never seen anything like it. All around us the silhouette of mountains and trees. In front of us are loads of rocks and with the reflection of the night sea and stars in our eyes, we settle down wondering what tomorrow will bring and hope we have camped safe.

MAGGIE:

When we wake up, we hear rain, but by 9.30 the sky is a brilliant blue.

We could not believe our eyes. Right there, in the front of us, dolphins were frolicking in the waves. It felt like they were putting a show on for us. Going upright on their tails and riding the waves. We found out later they were actually forming a hunting circle. We felt so privileged to see such a spectacle.

We sat and watched them for ages. I don't know what it is with dolphins. They just have this ability to make us humans smile. Even thinking about them melts the heart. Maybe it's because they look like they are smiling at us and we can't help but respond.

When we went for a walk to the nearby rocks, we saw three large seals basking nearby and lots surfing in the distance. We spend the day basking like the seals. This camping area is fantastic and we have it to ourselves because of the time of year.

Later, we got to see a seal close up. We just came upon it. It just lay and looked at us with its beautiful big, soulful eyes. It has a face like a bear and has brown fur too. It's flippers are in the same position as the Dracula like pose of the Fruit bats we saw hanging from the trees, way back in the land of Oz.

PETE:

We wake up and we have camped safe alright. When we open the door, there, in front of us, are dolphins jumping out of the water. That's all it takes for us to make our minds up to stay at this site for as long as we can. It is really beautiful here. Exactly what it says on the tin. The mountains are luscious, the water very clear, the landscape itself is stunning and we have been told that New Zealand just keeps getting better at every turn on the road.

We walk along the beach and the sandflies are out in force. It's not spoiling my adventure (I refuse to let it) but, I am being bitten continually whereas Margaret gets left alone.

Something caught our eye and as we go closer we see it is a huge, brown fur seal. Margaret goes over to it, standing next to it on the adjoining rock. It's not until later on, we see a poster in town warning, 'Please give seals a wide berth, about ten metres, as they can be aggressive'. Margaret wisely doesn't go as close again and neither do I.

It's another starry, starry night. We get settled and hope to see the dolphins again tomorrow morning.

MAGGIE:

The weather has turned a bit cooler and we need jeans and a fleece for the first time since leaving Scotland. We saw the dolphins doing their thing again this morning. Folk would pay money to see such a delightful display.

We eventually drive into town. The rail track is right next to the road and we managed to get in synch with the train for a wee while. Felt like Casey Jones was waving at us or was he shaking his fist? The train slowed down. We drove on and stopped further up at a lay-by and got out the van. As the train passed, the driver tooted and waved at us. Yep, it must be Casey Jones.

We decided to book a whale trip for the next day. We were offered a plane trip for the same price but we have a good feeling about the boat trip. They have guaranteed if you don't see a whale, you get your money back. Seems a very fair deal.

PETE:

Last night the stars felt so close. Honestly, it is a dream come true seeing nature and the stars in all their glory. Why anyone would not enjoy camping I'll never understand? We can't wait to open the door to see if the dolphins are there. As I open the door I notice that another van has stayed the night. For some reason, I feel like telling them to bugger off, which is unfair really as there is room for a few more campers. They probably just want to see something as well. I bite my tongue, open the door and go out. Yep we are lucky. They are there again.

Later, we drive into Kaikoura where we discover there are boat trips that take you whale spotting. Margaret is in the office before I can say 'there she blows' and before I know it, we are booked on a whale sightseeing trip tomorrow. I wish she'd blow on my legs and arms as I am well and truly bitten all over.

We enjoy the rest of the day exploring the town and discover a Brewery with four different ales. We get to taste them all. They are all good, but our favourite is called Run for the Hills. "It's what you do if a Tsunami is coming," the woman explains. To be honest I don't know if I could Run for the Hills after a few of these, in fact I don't know if I would be caring, as the ale is quite strong. We decide to take the risk. Back home on the Isle of Arran they have a real ale there which is great and hoppy called Red Squirrel. Arran is the only place you can get it. It's a pity they don't have that. Margaret says it is her favourite.

We also go for a walk and come across an archway made of whale jaw bones. They are huge. Quite a sight. You feel sorry for them, but they are from years ago and thank God they don't hunt them anymore. It is a haunting reminder of how much they were hunted in the past. So glad we have a chance to see a real live whale tomorrow.

Whale of a Time

MAGGIE:

We are so looking forward to our whale watching trip. When the time comes, we were proved right to go by boat. There must have been at least one hundred dolphins flipping in and out the water, doing backward flips and swimming next to us. They did their magic again just like the kookaburras in Australia. It was so uplifting, my heart felt as if it was having a fiesta.

We then went quite a bit further out to sea and were rewarded. Tariki is a male Sperm Whale, who is as big as our catamaran. He is surfacing and spouting water like a fountain. Seemingly he lives a solitary life until mating season. The ladies are a third smaller than the adult males and need a hotter climate. He goes and does his stuff then needs the cooler climes of Kaikoura. Thank you Tariki for forgiving us humans our past and letting us meet you.

As the catamaran turns I spot a seal with its flipper raised up out of the water, then it's head came up just as we went past. Felt it was giving me a wave so I waved back. Very special day.

PETE:

Here we are in Kaikoura. Hearts are a beating loud, as we board the boat, in the hope of seeing our very own Moby Dick. It's a fantastic day, the sun is shining, the waters are turquoise blue and all we need are dolphins swimming along the side of the boat like you see in films.

We are just a short distance out and all of a sudden there they are! Dolphins swimming alongside the boat like you see in films. Loads of them. We can nearly touch them as they leap and swim beside us. Then there's a shout 'MORE DOLPHINS', and we look outwards there seems hundreds of them. Some are leaping, some are backflipping, some are doing some really impressive underwater stuff like backward wheelies and even synchronised moonwalking. Really showing off. Only, we find out they are not showing off. They are actually hunting and are in fact surrounding what must be a huge shoal of fish. The splashing is telling each other positions. MAN! what a sight!.

As we sail on, we get told sonar has picked up a sperm whale: one of the biggest mammals on Earth. We were told it will come up for air but only for a short time as they dive up to two miles or more down. I mean 'two miles' that is a long way down. They also hold their breath for up to two hours (try holding ours for one minute) We don't want to miss this so we get as close to the edge of the boat as we can without falling over.

Out of the blue, there she blows. I actually get to say that to myself. It really is there he blows as the male sperm whale blows air out the top of its huge head. It is absolutely huge. Bigger than the boat we are on. Everyone is taking photos and then the woman announces, "Get Ready, it's going to dive back down and now's the time to see its massive tail out of the water before it disappears. The tail appears and you are in awe of this. Everybody should get a chance to see this. This magnificent creature."

I also learn that they sleep straight up and down like posts in the ground. I feel so privileged and lucky today and we got to see so many dolphins hunting as well which was a bonus.

Back on land I see two butter fish although they were battered and in a box with chips.

MAGGIE:

We got speaking to a couple of young girls from Switzerland. We realise how lucky we have been, since they told us they went on the same boat trip yesterday and saw absolutely nothing. They got their money back, but what we saw was priceless. You could see that the girls were so disappointed but, then there eyes lit up as they told us about the seal nursery just up the road.

Needless to say we are now going to stop at Ohau Point to see the baby seal nursery and we offered to give the girls a lift up. One good turn deserves another. What a find. I didn't expect it to be actually like a nursery. About four adults are dotted about, supervising and yes there they are: lots and lots of baby seals diving, swimming, play fighting and generally have lots of seal fun. I could have stayed and watched for hours, but the light was going which forced us to leave.

We thank the Swiss girls so much for this tip. They have managed to get a lift back to Kaikoura so we say farewell as we are heading up the road towards Blenheim.

PETE:

Later that afternoon we visit a seal nursery we heard about. It has luckily survived the earthquake as the landscape has been changed forever. It was really cool looking. We can't go too near. I don't know if there are many seal nurseries in the world but this is a natural thing. A few adult seals are sitting on various rocks and the little ones all frolic about in the pools and rocks caring not a jot.

MAGGIE:

We ended up driving and driving. The 2016 earthquake means there are roadworks everywhere and therefore not many stopping places. I noticed at Kaikoura that there were lots of women working with the roadworks and that many of them were Māori. They were the cheeriest road workers I have ever encountered, smiling and waving good naturedly as we drive past. They seem to have managed to preserve their own language. In fact, Kaikoura, which translates to mean crayfish food, had a very hippie vibe about it and I could easily have stayed there a lot longer, but we want to travel all over and we don't have the time.

The roads are also very, very busy with huge trucks. We notice there seems to be more road rage about up here. So we kept on trucking because we had to. There was literally nowhere to stop, until we came across a parking area in the middle of nowhere. There were four or five cars and campervans already there and a portaloo. It is permitted to stay overnight in these designated areas and it is free. We fell asleep pretty quickly.

PETE:

As it gets dark we move on and discover it is a long, long drive with no stops anywhere. Eventually we come across a parking place off the road and we draw in. It has a toilet so we are allowed to stay the night. We discover the continuous traffic of the huge trucks makes it very noisy all night, but I manage to sleep on and off. When I wake up, I discover that my lotion has worked better this time, but still only for a bit at a time. I am still looking for the missing ingredient.

We look at our trusty map and head for Blenheim and after that, towards the McKenzie Mountain Pass which will allow us to loop back to Christchurch and then eventually towards the Pellanor Fields location home of the Riders of Rohan from Lord of the Rings.

MAGGIE:

Arrived at Blenheim. Pete suggests, as it's a beautiful day, let's sit and enjoy the nice weather. There are a few ducks on the river. No exotic wildlife on New Zealand's inland, just in their ocean. We have found out that when white settlers arrived on New Zealand the only native mammal was a small bat. There were lots of birds and because they had absolutely no predators they had no need to fly. The white settlers decided that since they liked hunting for the sport they introduced rabbits. Again they had no predators and bred in their millions. So in their ignorance they introduced stoats in their belief that they would control rabbit numbers. It was much easier for the stoats to eat birds eggs and so they preceded to make forty eight species of native non flying birds extinct. The Kiwi, the national emblem, survived only just and is on the endangered list.

We stayed at a campsite for truckers and I didn't like the vibe. The sullen women had no time for us. I was glad to move on.

PETE:

Here we are in Blenheim and I must say that it's not what I expected. A bit drab, but we found a Truckers campsite. We find a spot and I head to the little shop/cafe. There's a bunch of truckers, all big guys, and they are drinking beer and swapping stories. Trying to outdo each other with who drove the up closest to the guy in front's arse. Well! I can vouch for the guy who drove behind me. He wins. I couldn't even see his windscreen in my rear view mirror, just the grill of his truck.

This place reminds me of a Route 66 type of place. All the trucks here have at least thirty two wheels on them: sixteen on each side. Plus they are full of timber or suchlike. Usually not a good feeling while driving bumper to bumper with you. Anyway, rather them than me I suppose. They have earned their beer for the night I reckon.

Our reward for the long drive was to see the Pacific coast steam train again. As the name suggests it only runs along the Pacific coastline and as usual you get a toot toot and a wave from the driver and sometimes the passengers. It sure does bring a childish smile to your face, unlike the Formula One truck drivers.

MAGGIE:

We don't really want to repeat last night when we drove for hours in the dark and saw nothing. We are heading into the Wairau Valley which is a huge wine region. The place we are heading has a French name Saint Arnaud and that seems appropriate. We stop and start as we drive along enjoying the view. When we arrive at Saint Arnoud, we decide to drive up to Nelson Lakes National Park where the misty snow-capped mountains reminded Pete of a Led Zeppelin song.

It was very blustery, so I admired the view from the comfort of Noah the campervan. Pete did venture out to the boardwalk and looked windswept and interesting. He saw really big eels in the water.

We keep driving further up into the mountains towards the pass. The weather starts to improve dramatically and blues skies appear. We stop at an interesting looking wee church up in the mountains looking onto a lake.

It's sign says: Come and Worship the Living God. I love that message. As we go into the church I see what they mean. There are huge windows overlooking the stunning view. The Living God of Nature is all around us and we sit and soak up all its splendour. Then we give thanks and remember to count our blessings: for our good health and being able to live our dream of travelling.

Up, Up and Away

PETE:

After Blenheim, we discover that they are not kidding about the scenery. Photos don't do it justice. There are fields and fields of grapes. I wonder, not for the first time, how they manage to pick them. It's a great sight to behold and hopefully just as great to drink, as I am sure we will find out.

As we drive on into the hills, it feels like real New Zealand Hobbity stuff: all misty mountains and craggy cliffs. To help us on our way we decide to have some Run for the Hills Real Ale later. The stuff we had at Kaikoura. It is quite apt we think. We have been here nearly a week already and I hope it doesn't fly in like Australia. Talking about flies. I sure am fed up of being bitten and staying in truckers paradise last night did not help.

Here we are heading up the McKenzie Pass and it's up, up and even more up. Brilliant scenery, but narrow, twisty roads. We turn another of the countless twists and turns of the mountain and right there, in the middle of no-where, in front of us, is a pub and a car park. We decide to stop as Margaret's brake foot needs a rest.

I notice the name of the pub is the name of my big brother. I wonder if it is Welsh owner and it turns out that they are in fact Asian. We have a coffee and I'm surprised to see a Mrs MacPie which reminds me of Australia. I notice a juke box.

I can tell a lot about a pub by its jukebox. I notice that there are three credits left on it so it can't be much good. I press 09 for the only decent song on it. Moon Shadow by Cat Stevens. As I sat down, a song I had never heard of came on. I double checked 09 and definitely says Cat Stevens Moon shadow. No wonder there are credits left. We decide to leave and I leave the remaining two credits for the next passer-by.

We eventually see a signpost that says Shenandoah. I immediately take a photo for my twin brother as that is his favourite cowboy thing. I must mention that I miss the Australian signposts. They were hilarious. I mean: Butt Creek, Blue Knob Creek and Little Tilly Willy Bridge to name but a few. I miss the Australian humour. I could do with some of their insect repellent we saw called Bugger Off. As we pass Shenandoah we see another sign that says: Welcome to the Biggest Swing Bridge in New Zealand. We know that the New Zealanders like an adrenaline rush, so this we must see.

As we arrive the guy at the box office says we close in thirty minutes, but for five dollars we can go in and if we hurry up we can cross the bridge. We go onto the bridge and it's very swingy. Reminds me of the swing park days when you worked up a swing sideways while standing up. Only this was very long. Very high. We ended up jumping up and down on it, but not through choice. It was so swingy you couldn't stop it being swingy. We only managed halfway over and as the guy wanted to close the entrance, off we hobble. We are so glad to be on solid ground that we sit for a while.

MAGGIE:

When we come across a sign for the longest swing bridge in New Zealand, we decide to investigate. It is apparent that New Zealanders have a love of high adrenaline pursuits. After all, in 1986, A J Hackett invented the bungee jump at Greenhithe Bridge, Auckland. Maybe it is because it is a land of extremes. Volcanos, glaciers and earthquakes always a real threat. Probably, they need something to use up the adrenaline that it is there anyway. They don't call it fight or flight for nothing.

The guy at the entrance explains we have only half an hour. We go for it and as we turn the corner take a sharp intake of breath. This bridge is one hundred and ten metres long, very narrow and very, very high. We go on. It does not feel very secure. I can see Pete's fear probably mirroring my own. I manage to get out halfway and then see a young couple on the other side wanting to come back over. We happily came back as we felt it was only good manners to let them get home.

Speaking of adrenaline it makes me think about my own addiction to it. It is not so obvious. I don't enjoy bungee jumping, walking across one hundred and ten metre swing bridges or any extreme sports. It is more subtle, but it is still there in my everyday habits. Running late all the time like the white rabbit. Even when I am on time for something I somehow manage to find a distraction just so I am cutting it neat. I don't like being late so that is where the adrenaline comes in. I put myself under pressure (I didn't even realise and it was subconscious until recently) just so I can cause an excuse to blame myself and internally beat myself up therefore causing an adrenaline rush of anxiety. Leaving things to the last minute is really just one example of covert adrenaline addiction. Such a habit. I am determined to kick the habit and live a more peaceful life.

PETE:

When we get our senses back and we are back on the road again, we stop to look at a waterfall. We come across a guy with a tripod and his camera pointing at the waterfall. A guy in a canoe suddenly comes into view at the top of the waterfall and he heads right over it head first. After what seems a while, he surfaces. Then he comes out, runs up to the camera and it's the turn of his friend to sail over. He waves as he is doing it. Another one of those adrenaline New Zealand hobbies.

I then spot a noticeboard which explains that this was once just a flowing river, but because of the earthquake three years ago part of the river bed slid down eight metres creating the waterfall. Yet, another example that New Zealanders just get on with it.

I can't stop thinking about earthquakes for a while and get quite edgy as could happen again quite easily. However, the feeling soon passes as we turn the corner into another stunning place which looks like a cross between Switzerland and Scotland as well as New Zealand. It is jaw dropping when we wake up, see it in the early morning mist then turning into glorious sunshine. We go for a walk and end up on a forest track and come across peculiar looking things. Man made things, but we can't make out what they are.

MAGGIE:

Our surroundings inspire us to go for a walk and we see evidence of a fault line from years ago. We also see a noticeboard with various native birds. I notice a picture of a Kea and also the rarer Kaka both of which look like a cross between a parrot's head with body of a hawk. As we walk along a charming little bird comes up really close to us on a branch and starts a displaying its tail. I recognise it from the noticeboard as a fantail. We will see a lot more of these lovely wee birds as we find out they are quite tame. Maybe it's because they love to show off their beautiful fantails so much.

We are going to head to Hanmer Springs which is about an hour from Christchurch. We are going to try out the hot sulphur springs there. A couple of miles up the road we come to a hot springs Hotel and decide to stop for a closer look. As soon as the girl at reception hears our Scottish accents she starts by saying it is officially the sexiest language in the world. This is news to us. I thought it was French I say. Oh no she says Jamie from Outlander put a stop to that. Turns out she is from Austria and travelling around for a while, working here and there. Her English is fantastic.

After a plate of soup, we decide to move on as Pete gets a feeling it's just a tourist trap. He is right enough as the hot springs at Hanmer are much cheaper and look miles better. It is just what we need, especially for Pete with his bites. There are various pools and the water is deliciously hot and steamy. With the cooler air on our heads, it is really invigorating.

PETE:

We are heading to Hanmer Hot Springs which is a long drive down the road. We are quite surprised to come across a big hotel advertising hot springs just a few miles down the road.

As we stop and go in I immediately don't like it. It feels too touristy for me. I am not hungry, but Margaret fancies some soup. When it comes, it is not good value. I guessed correctly. As I am waiting on Margaret I get talking to the girl behind the counter. She comments that she likes my Led Zeppelin t shirt. I tell her I saw Led Zeppelin in 1979 in what was to be their final tour because drummer John Bonham died the following year. Her eyes go as wide as saucers.

Then she says she likes Black Sabbath. So do I, in fact, I tell her I have seen them as well. Lucky you she says I was supposed to see them in Christchurch just a few weeks ago, but it got cancelled at the last minute. Why was that I ask ? The support act were Slayer, but that guy sprayed all those people with bullets the day before the concert and it was deemed not appropriate. Yeah we were in Australia when it happened. Real sad. I remember a few Australians saying the felt uneasy because it was an Australian who carried out the attack.

We move on and finally, after a wee bit of a squeaky bum drive, we make it to Hanmer Springs.

It certainly is not as expensive as the other one. I think we got in for the price of the soup. Once in, we went round a few of the pools, each one a different degree of heat. I feel as if the hot water is helping my bites. After sitting in a few of the pools, Margaret heads towards the sulphur pool. We go in and it's stinking, pure sulphur. I then realise as I look at Margaret's face she is over the moon. Her soup has started to take effect and no-one notices. She is in farting heaven and getting away with it. She's not the only one I bet. Nobody is any the wiser except me. I see the bubbles.

MAGGIE:

We stop at a beautiful beach and we realise that we have come full circle from last week. The first campsite we stopped at, as we arrived, is just down the road. Tomorrow we are heading to Christchurch and as we have done for all the big cities, we are travelling into the city by public transport. Before we head to Christchurch, we have our breakfast at the beach, sitting on a huge piece of driftwood, specially sculpted by the sea.

When we head down towards Christchurch, we are popping in to the campervan hire company to collect some extra blankets, then parking in the suburbs. We eventually jump on a bus and the driver, when he hears our accent, informs us he has been to Britain. What do we think of our fish suppers here? I love the fish but the chips are very disappointing as we have only had frozen chips both in Australia and New Zealand. The UK really does know how to make good chips. The fish here, however, more than makes up for it. The choice is fantastic. I especially love blue cod. Pete loves Barramundi best.

PETE:

Out we come, fresh as daisies and we sleep like babies. In the morning, we go for a nice walk. The place looks like a French or Swiss ski resort. Braw and kind of Christmassy even although it's April.

As we head away down the road, it's another warm, sunny, autumnal day. We see a cloud of black smoke, quite thick, in the distance. Oh no! we think in unison-it's a fire. What do we do? Fires can start in New Zealand just like Australia and the damage is awful. We slow down as we get closer. The smoke is really black. We hear an engine noise and we pull over behind a rock to let, what we think is a fire engine pass. In front of us, the noise gets louder, the smoke gets thicker and round the corner comes a steam train putt, putt, putting and chugging for all it is worth. We all wave at each other and get a toot toot. We just love the trains.

We keep driving through the very scenic island hills, full of poplar trees and find ourselves at Amberley again. We stop at the beach for a while. I find a starfish and a closed shellfish like a pearl clam. I open it and there is a small, black thing in it and for a split second I think it could actually be a rare black pearl. It's not-it's a very small black pebble. In fact, the shell wasn't closed. It was open, but with the two sides resting on top of each other. At least, the starfish was real, albeit dead.

Once we pick up the extra blankets we decide to get the bus into Christchurch city as it's a big busy place. Sadly my bus pass doesn't count here in New Zealand. As we walk around the city there is a sad feeling or vibe, Most people, understandably, are not over the shock of the mass shooting here a few weeks ago.

We head for the real ale bar which allegedly sells most ales. We are disappointed that Run for the Hills has sold out. It always does says the barman so we settle for something else from the same brewery in Kaikoura. It is nowhere near as hoppie. I tell Margaret that I am going to do some cosmic ordering and 'put it out there' I want our favourite ale 'red squirrel'.

Christchurch at night all lit up is really quite nice and we do a bit of sightseeing, but because we've come by bus we have to watch our time. By the time we get back to Noah, it is really quite late and dark so we drive to a camping area and go straight to bed.

Sometime later, I get out to go to the toilet and I see something at the van door. I bend down to pick it up and for the life of me it's a small, plastic red squirrel. We've still got it. It sits on our dashboard to remind us to be a bit more precise in what you ask for.

MAGGIE:

To be honest I am not that keen on Christchurch. It's OK as far as cities go and if you like shops. It has had a lot to put up with in recent years between the Earthquake and the shootings. There is a definite sadness hanging around which will take some time to shake off.

We decide to try and see if we can find our favourite New Zealand ale Run for the Hills and head to a pub which boasts having a huge range including our favourite. We are out of luck and drown our sorrows by having some Enigmatic ale which is good and from the same brewery. Pete decides he is going to do some cosmic ordering for some Red Squirrel Ale which is brewed on the isle of Arran back home. This will be really difficult because it is only sold on the island of Arran.

Back at the campervan, after we have moved to a suitable camping area, Pete nips out and I am astounded when he comes back. He has found something on the ground. In his hand, he has a tiny, wee plastic red squirrel. I laugh like a kookaburra. I didn't think he had a chance of finding any red squirrel in New Zealand. How wrong was I?

PETE:

We wake up to another warm day. We feel pretty good on our travels and both sing 'Life's been good to us so far' at the top of our voices.

On the road, we pass a roadside sign saying real Manuka honey. This stuff is straight from the hives and cheap considering Manuka honey only comes from

New Zealand, it's world famous and very expensive back home. We get some and it tastes amazing. What a bargain since it has so many medicinal compounds.

We also buy some home-made soap. I don't know what Margaret bought, but I bought Animalistic to try and bring the animal out in me. Watch this space...

It's not that long until we come across a town called Little River. It has a real cowboy town feel to it. In fact, you could say it's a one horse town with an old train station and also a Cadet Centre. I heard that only recently, one of the girl cadets, through her training, managed to raise the alarm and stop a major fire. She spotted the signs of a fire. She saw smoke was under a pylon. Seemingly minute sparks come from pylons and because the grass is so dry it can flare up very easily. So well done young Cadet for paying attention.

MAGGIE

We are heading to the Banks Peninsula. Our lovely Wedding Celebrant in Burleigh Heads, told us that Akaroa is her favourite place to visit in New Zealand. French colonials settled there and it has a distinct French feel to it. It also has hector dolphins which are quite rare. It is a beautiful drive over and we see a sign at a farm on the way advertising Manuka honey. We feel we must try the native Manuka honey whilst here. It is made from tea tree flowers which only bloom in New Zealand and has amazing healing properties.

We walk around picturesque Akaroa, do a bit of dolphin spotting at the harbour. We think we see a fin for a split second. We already knew we were lucky with the dolphins at Kaikoura and this emphasises this fact. France is a favourite holiday destination of mine and so I recommend Pete trying some French Onion Soup later.

We have a surreal experience peeking into a strange wonderland much like Alice peeking through the looking glass: music playing Edith Piaf, giants in berets dotted around in various poses playing giant musical instruments, mosaic furniture, but no mad hatters in sight. It looks deserted. We don't want to tempt fate or drink anything, so we leave pronto.

PETE:

We come to a place with one of the most impressive views yet. We are high up, looking down on a place called Akaroa. It's a French styled town. Down we go, grab a parking place and go and explore.

We climb a steep hill and come across a teapot. We can hear French style music drifting towards us. We can't see anyone about as we venture up the steps and the music is getting closer. Next thing we know we are looking into a garden where things are giant size. We feel like we are in Alice in wonderland. It's very surreal with a giant playing a huge double bass and all sorts of weird things. As we are taking photos at the entrance it dawns on us that it is getting late and without much ado we head back down the steps.

As we head back, Margaret spots a shop selling merino wool jackets, hats and stuff, so in she goes. We discover that in fact the wool has possum fur blended in. As I imagined the sheep being sheared, I thought poor possums. We'd already seen lots of dead ones on the roads.

The woman explains that the fur is super warm. She also explains that killing possums for fur is an extremely good thing as there are still thirty five million of them. "Eh? What did you say?" "Yep thirty five million. There used to be around seventy million, but we're managing to get rid of them." I'm gobsmacked. I can't take it in that there are so many of them on such a small island. Besides aren't they the cute little things we saw in Australia? They shoot them and set traps on trees as well. So that's what we saw on the trees. The mystery is solved.

"Do you like French Onion soup?" asks Margaret. "Don't know," I reply.

Since we are in a 'French' town I'll treat you. We find a traditional French restaurant and order in my best accent le french oonion sooop sil vous plais. It's the best soup I have tasted in years with melted cheese on top and croutons. I didn't know what to expect, but I wasn't expecting that. Then she orders crème brulee which I have never heard of, but it tastes great. I feel spoilt.

MAGGIE:

I nearly bought a merino wool/blended possum fur jacket. I already have blankets back home made of Icelandic sheep wool and I know how great these are. They kept me very cosy back home in the campervan. They always keep the right temperature for body heat so they are ideal in winter and summer. So when the lady in the shop tells me about the merino wool and possum fur I know it is not hard sell. She is telling the truth.

I knew when I kept seeing dead possum roadkill everywhere that there must be a lot of them going about. I was astounded to learn that at one point there were over seventy million of them. Again this is due to the fact they were taken from their native Australia and brought over for hunting purposes. Again they didn't have any predators. They are vegetarians, eating the buds off native trees. Not only are the trees suffering, but also all the native birds are suffering. The numbers are OK in Australia. As soon as humans tamper with that balance of nature such as introducing a non-native species, they are asking for trouble. New Zealand is living proof of this. What a horrendous story, and there are still around thirty five million of them and the poor things are hated and persecuted here.

PETE:

We walk along the pier in the moonlight in this French town in New Zealand and then we say au revoir and drive on towards to a campsite at Little Akaloa. It's further than we thought and it is dark by now. The road keeps going up and up and this makes it feel even more remote and it's getting very, very foggy.

After what seems like an age, we see a sign for the turning to Little Akaloa. We spot two large eyes in the headlights. I get out and there is another one with its back to me, walking down the road. As I get closer it turns round and looks at me with these large eyes, petrified. Possums. Our first live ones in New Zealand. It starts walking away, like Pepe le Pugh the French Skunk. It is to become the first of many live possum sightings rather than roadkill: all in the dark as they are nocturnal. It's late when we reach the campsite and it's straight to bed again.

MAGGIE:

I do agree that the Italians have the best food in the world, but I have to say, for me, French cuisine is a very close second. I absolutely love France. I have been a few times, camping as a family with my daughters. Me and Pete hope to tour it in the future. Meanwhile there is a traditional French restaurant in town and so I am going to show Pete the sort of food he can look forward to when we do go.

Afterwards we head for the campsite in Little Akaloa which ends up being more difficult to find than we anticipated, probably because it is pitch dark by now. The law of gravity comes into play as we have driven upwards for miles. What goes up must come down and down we go. Down, down, deeper and down.

My knuckles are white. I am in first gear and it still feels too fast. Eeeek— as we turn a corner there is something on the road. I manage to stop and Pete jumps out. In the headlights, there are a couple of possums and one doesn't see Pete. It is waddling up the road at a leisurely pace. When it becomes aware of

Pete, it literally jumps up petrified on the spot, then speed walks away as fast as it can, but still swinging it's hips. I can't help it. It's like a cartoon character and I think they are very cute. I do understand that numbers have to be kept down and also I know how I feel about grey squirrels in Scotland. What a nightmare. For the possums and for New Zealand's native species.

PETE:

Next morning we discover that Little Akaloa is a lovely wee place with nothing much in it. There is a big tree next to us, so we decide to give our clothes an airing. I tie a line onto the tree and onto the back of the van and peg out the clothes. While they are airing, we go for a walk and come across a small beach with lovely views.

I find some Paua Paua shells. You only get them in New Zealand. They have amazing colours on the inside and once polished, also on the outside. We have seen them being sold either whole or made into jewellery. I take three of them that are just lying washed up on the beach like all the other shells.

We decide to explore further as we saw a sign for a Māori Church that points to an old track and which seems to go up through woods. The track is called Luke's track. It's quite narrow, a bit overgrown, but there are old steps leading the way. Margaret's bird senses kick in and she announces there are birds around here and sure enough wee fantails appear. They are really braw and if they like you they will show off and fan their tails for you. We then come to a sign which says Pied Shag Colony and as we go to look for the birds, we come across an old bunker of sorts which seemingly, was used during the War for the home front defences.

The Church appears as we come to a clearing. We try the door and it's open, but no-one else is there. We are met with impressive hand carvings of Māori designs with lots of Paua Paua shells inserted. There are also two big brass plates on the wall commemorating all the young men who died in two World Wars. I didn't even know that New Zealand were involved with the war. As I look at it I see my date of birth 6th May and the name of a solider but it's his date of death. A lot of families lost more than one son. It is a stark reminder of how many young people lost their lives. I need to go, but before I do, I say a prayer of thanks and once again admire the work of those who built the Church.

MAGGIE:

This is a little gem of a campsite. We have the place to ourselves. The other van, which was there when we arrived late last night, has departed. When we booked our trip, we landed just as the school holidays were over at the end of January. It is an ideal to time to travel over here. Not too crowded, the weather is not too hot and it is cheaper as well.

On the beach, we come across some Paua Paua shells. They have a sort of tough, calcium outer layer, but the amazing turquoise and green colours inside. They are so beautiful, I didn't believe they were real when I first saw them in the shops. I thought they had been hand decorated. Later I found out that they are a type of sea snail which they farm over here for their meat and also their shells. We have seen them on the menu in fish and chip shops, but we weren't brave enough to try one.

We manage to find an amazing wee church with many Māori influences. The stain glass windows have Māori patterns and there are interesting carvings all over. It had a great vibe. I have not had the chance to go to any New Zealand church services yet. As I have already said my church is wherever I am.

The solitude of the van has given us the opportunity to be close to nature and to also be close to our own true self. Keeping it real. Being in a state of calm reflection. I recommend it for everyone. The world would be a far better place for it. Instead of masking hurts, frustrations, disappointments and anger behind noise and busyness, own up to them and let them go. Silence is golden.

Pete notices a plaque on the wall in memory of a soldier and something touches him. He died in Tunisia on 6th May 1943. Pete was born on 6th May 1955. It touches him profoundly.

PETE:

As we head over the hills back towards the end of the peninsula we stop at Okains Bay to look at the best Māori Museum in New Zealand. As we go in there are loads of carvings on the wooden buildings with sticky out tongues. This is Tiki protection.

There are lots of large pictures and paintings of past Māori Chiefs and Māori women. I notice that the women have tattoos on their faces as well as the men. These are clan markings. We find out that a woman would get all her tattoos done at once whereas men would get their faces done one step at a time. They are not wimps. Far from it as they get the hairs on their face pulled out one by one so they can't grow back in. OUCH. So next time you meet a Māori Chief, show off and ask him if he's had his beard pulled out instead of shaved.

We head across to a shed where I spotted a guy working in a canoe or Waka as it is known in Māori. We find out he spends most of his time carving. He takes a shine to us as he tells us he has Scottish ancestors. He is actually Canadian and we have a real good talk. He invites me up into his waka to get my photo taken. He told us the story of how he was with friends out at sea when they got lost. They just lay their waiting, hoping. All of a sudden the clouds disappeared and the stars appeared. They recognised the southern cross, a formation of stars which acts like a compass. If you sail towards the southern cross, you get to dry land and it saved their lives. Nice story.

MAGGIE:

We are in Okains Bay, a town we had no idea we would stop at until we notice a sign for a Māori Museum. We spent a few hours wandering round it. It was a collection made by one man's passion for the Māori culture. His name was Murray Thacker and he was white or Pakeha in Māori language.

One thing I noticed was that quite a lot of Māori women married white settlers especially Scottish settlers. The Māori have total belief in the great spirit but also the need for protection against the dark energy. The Tikis outside the buildings are an example of this. Their tongues sticking out in a HAKA expression frightening the bad spirits away.

I totally understand the theory that everything, absolutely everything, on this earth has energy, some light, some dark and this goes for humans too. We have been given the free will to choose which of the two paths to take.

I remember seeing the programme Blue Planet. Even in the deepest of oceans there, there were fish down there that looked angelic and some looked evil, like Orcs from Lord of the Rings. Speaking of which we are heading to McKenzie pass and there are lots of Lord Of The Rings locations around that area and then heading down towards the Fjordlands.

PETE:

We're off to find Ben Ohau going through McKenzie pass. To us Lord of the Ring fans it is better known as the Pelennor Fields, home of the Riders of Rohan.

It takes longer and is further than we anticipated and so we stop at a town called Geraldine to stretch our legs. We decide to go to Talbot Scenic Reserve for a walk in the woods. We notice that it has totara trees and we read that these need to be protected, like everything else native to New Zealand does.

We drive on it's getting darker and darker and wetter and wetter. This is the worst rain we have had since Australia. We can hardly see. The window wipers are at full blast. We keep on driving for what seems like an age. Getting lost in a campervan is half the fun, but when it's the other half it's not so much fun, but that's what it's all about, so we keep going ever further. To our delight the rain starts to ease off. What happens next takes us completely by surprise.

All of a sudden there are clouds in front of us, but not your normal clouds. These are masses of moths. Millions of them. It's like a plague of locusts. When I say millions, I am not exaggerating. They just appeared after the rain and they are everywhere. It is quite spooky in a sense. At last, we find a campsite here just at the edge of the McKenzie mountains. The owners said to take a long lie-in if we wanted. This turns out to be a Godsend. Most campsites want you out by 10 am.

When we get our pitch, I open the door and I immediately wish I didn't. Attracted by the lights the moths come whizzing in. It takes all night to get them out one by one. We are afraid to open the door in case another batch come in. Not a good night. In fact, it was late when we got to sleep and we didn't get up until noon.

MAGGIE:

It is raining most of the day. When I think about it, I realise that we have hardly had any rainy days. A couple of electrical storms at night and the odd shower. In fact, I can count on one hand how many rainy days we have had. Coming from Scotland this seems miraculous.

All of us on planet Earth are at the mercy of the weather. How we deal with what is thrown at us is what makes us. Australia and New Zealand live on the edge of extremities all the time.

We end up at Geraldine and walk round the town for two or three hours. We come across a St Andrews Church. I notice the stain glass window has Jesus holding a lamb. I had just been talking to Pete about his take on the symbolism of the lamb. He said suffer the little children who come unto me for they shall inherit the Earth. I think I am beginning to realise just what this means. Every one of us is made to suffer and it brings us to realisation that Love is all that really matters.

PETE:

On the way to Ben Ohua, we come across Lake Tepako and stop to admire the view. At the Information Centre at Lake Tekapo, we find out there is a space observation point alongside a space laboratory on Mt John. We definitely want to see this as it is seemingly one of the best places to observe the night sky.

As we head up I notice the grass is like a huge prairie going on as far as the eye can see. It is an eye catching yellow colour. It's great to walk around the observatory. There are loads of bus tourists and I can even see some monks.

Man, what a view. We decide to have a coffee in the Astro Cafe. When they arrive, they have the shape of Saturn on top. We sit and admire this for a moment. Just as I am about to drink it, a young Asian girl asks if she can take a photo of my coffee. I say yes and soon wish that I hadn't. Within the blink of an eye, a queue formed, all wanting a photo of my coffee. It was cold when finally I got to it, but it was famous.

On the road down the mountain, I spot the plant Mullein growing nearby. We pull over so I can collect some as I'm going to use it to make an oil for my ears. It helps to restore hearing and keep the ears in good health. Also drinking the leaves as a tea is good for chest infections.

MAGGIE:

Up till now the scenery has been beautiful, but as we come round the corner towards Lake Tepako, it steps up a notch. It is the purest, turquoise water I have ever seen with Mt John in the background and another small church to explore. The Church of the Good Shepherd ushers us in.

When we go up to the space observatory at Mt John, the scenery goes up another notch. I don't know how to describe it to do it justice. It is just so breathtakingly beautiful. First of all there is the huge prairies of yellow grass, then stunning turquoise blue of Lake Tepako and mountains everywhere you turn, the whole 360 degrees. I actually wept. I reckon my eyes were so happy to see such a sight, the colours are so vivid. You don't want to leave because you can't believe you will ever be lucky enough to see such beauty in nature again.

There are bus load upon bus load of tourists wanting to look as well. When we go into the café, it is busy. We just manage to get a table and we order two flat whites. They come with a chocolate covered Saturn on top. I thank the guy for taking the time to do this, when he is so busy. This is not busy he says, this is a quiet day.

PETE:

It takes a wee while to actually find the Ben Ohau campsite, but once we get our bearings we eventually find it. It is very popular spot for fishing. An ideal place as camp fires are allowed and it's a short walk away from the Fields of Pelennor.

When we wake, I don't think I will ever forget the sight: in front of us are eighteen beautiful, black swans gliding past us on the lake which looks like a mirror. Surrounding the lake are a ring of rolling mountains and at the bottom of the mountains are the Pelennor Fields.

As we walk there, I can easily visualise thousands of galloping horses thundering across the fields. I'm such a Lord of the Rings buff, I actually wear an engraved ring around my neck. Just like Frodo. It's more than a surprise when I look down and see the ring lying on the ground with the thong loose. I had tied it with three knots so how did this happen? "Maybe it doesn't want to be here," quips Margaret. Not funny I think to myself. I count myself very lucky to have glanced down at that moment to find it in the long, yellow grass. I have had the ring round my neck for years. It helps me to keep going when I feel like not. I'd hate to lose it.

I shrug the incident off as we look for a suitable spot for a picnic and a couple of beers. After a wee, while I go for a walk round the field to soak it all up, while Margaret enjoys the sun. I hear a distant call: "PETE." As I turn round there at the banking at the other end of the field, Margaret is waving something. I wonder what's up I say to myself. I head back and ask what's wrong. "Eh," she says, "when?"

"When I heard you calling my name and waving."

"Oh that," she says, "I got too warm so I took my bra off and thought I would let you know what you were missing LOL." I don't know whether to laugh or cry. I decide to laugh. We are having such a great time together.

MAGGIE:

We are camping at a river near Ben Ohua which was the site of the Pelennor Fields. This location was chosen because not only is it stunning (as is most of New Zealand), but it is vast and also near to Twizel for local amenities. Most of the people living in town at that time were extras in the film playing all sorts of orcs, elves, dwarves etc. What amazes me is that it is twenty years since Lord Of The Rings was made and still people like us want to see the locations. It is that epic. Pete has tried to explain why he loves the story so much. He says it's the triumph of good over evil, the value of true friendship and the burdens we have to sometimes carry…

We walk along the canal which is the same stunning turquoise colour of Lake Tepako. If I had to choose a favourite colour, then it is this. My eyes never tire looking at it and feels very healing. This means so much to Pete. He is indeed, my very own Frodo. He even wears an engraved ring round his neck. I realise that he has that simplistic good heartedness of Frodo. He is not in the least materialistic. He is wise, calm and good natured. He would be an ideal candidate to be trusted with the ring and not be corrupted by its power.

A strange thing happened at the Pelennor fields. The ring fell off. The ring had been round his neck for years, at least ten. It had never been off in all that time as it had a leather thong with three knots. Luckily, he just happened to notice it on the ground. Just after that two ladybirds landed on his hand. I took photos.

PETE:

As we reach our van, Margaret says, "I can feel a fire coming on." I agree. It's such a nice, calm night. So I gather the wood and create a circle of stones next to the river. The black swans are still there. We light the fire, grab a beer, watch the swans. As night falls, I look up and my favourite line of my favourite song The Wizard by Uriah Heep comes into my head. How it came to be my favourite is another story.

MAGGIE

I have decided to treat each day like it's special because it is. We walked back to the Fields of Pelennor/ Rohan today making the most of this magical place. Pete is going to wash his face in the river. The black swans have gone this morning. We had a wee campfire last night celebrating the full moon energy, the stars, nature.

Time to head to Mount Cook. We only decided last night to take the hour drive up to see what it's like. The nearest petrol station for unleaded is Omaramah and I can always nip in and see if there is any Run for the Hills ale. Turns out there is only one bottle left. I am saving it for my birthday tomorrow.

The drive up to Mount Cook is spectacular. It is actually difficult to describe just how spectacular it is. It feels like we are driving into a movie depicting heaven. I let my eyes feast on it and soak it up. When we make it to the campsite, it actually feels we are in the foothills of a mythical land.

PETE:

The swans have moved on. We feel so lucky to be at the right place at the right time. I can't stop thinking about my ring falling off. How did three knots get undone after all these years. We go for one last walk to the Pelennor fields and I wash my face in the river and drink in the scene as well as the water.

It's 14th April and we're on our way to the famous Mt Cook, the highest mountain on New Zealand's South Island. It's a straight road right up to the campsite at the foot of the mountain. We don't think anything can ever top the scenery of the past two days, so we are not expecting to be overwhelmed once again by Mother Nature.

Way in the distance we can see the peak of the mountain top above the rest. It's covered in snow. The mountain scenery all around us is very Lord of the Ringsy. All of a sudden to my left I see a huge animal.

"Hey," I shout, "did you see that! I just saw an Elk."

"What's that?" says Margaret, "did you say an Elk or an Elf?"

"An Elk I just saw an Elk."

"Well I feel as if I am in Middle Earth so you're bound to see an Elf. Actually I'd rather see an Elf than an Elk."

"OK. Very funny," I say.

"Whose joking," says Margaret driving on and looking for elves.

As we drive on it starts to get dusky and the sky is a very beautiful pink with hints of orange. The lake is like a mirror as if one thing is on top of another. There's also a huge full moon right above the road lighting up our way. If I did see an elf now, it certainly wouldn't surprise me.

The campsite is busy. We find a pitch and get our heads down for the night.

"I don't care what anybody says I remark to Margaret I saw an Elk."

"I know," she says, "and I saw an Elf!"

MAGGIE:

What a view to wake up to. Despite the snow on the mountain tops, there is not a cloud in sight. We find out later that it is very unusual to get good weather up at Mount Cook. We had breakfast in our glorious surroundings and it already feels like the best birthday ever. I decide to choose a Native Spirit Card. I got the

Eagle Medicine Card the same as I did at Burleigh Heads that special last day there, when my wishes came true.

Going for a walk to the Glacier, it turns out the bridge is closed due to path erosion. We take another route to the Mueller Glacier. We spent ages exploring the area. The water in the mountain lake, just off the Glacier looks milky and cloudy and then we see a strange phenomenon as the milky looking water evolves, mingling into crystal clear water pools right before our eyes.

Pete says he is treating me to a mudpack beauty treatment straight from the mineral mud all around me. Pete makes a slate scalpel and palette and proceeds to cover my face in this mineral mud which is fresh from the mountain lake. Then I relax in the sun and let it do its thing.

Pete then decides that we should do a HAKA face, but when I look at my photo it looks like I am letting the doctor examine my tongue. I think I need lessons. I try to put more 'don't mess with me' energy into it. A bit of an improvement. It's in the eyes.

PETE:

We wake up, look out the van and there is a huge snow-capped mountain right in front of us. It's 15th April and its Margaret's birthday today and what a beautiful day to have for it. I go for a wee walk. I feel so alive. The air is so fresh. It feels so clean here.

I keep hearing a noise though. That's the second time I've heard it. As I turn a bend the mountain comes into focus and see all these mineral pools. I decide to go back for Margaret to show her these. As I got talking to someone I heard the noise again.

"Avalanche," says the guy in answer to my look.

"A real one?" I ask.

"Well, erm, yeah a real one on the other side of Mount Cook. You don't get many this side, but hey you never know."

When I reach Margaret to tell her I found a place to go, she decides to get her Native Spirit Cards out. She loves them and I like them too. These cards are full of wisdom and always accurate. She draws a card. It has a snow-capped mountain on it and an eagle gliding over it. It's Eagle Medicine she exclaims knowingly. It means it time to soar like an eagle and look at the big picture...

"Would you like to come a walk, Oh wise Woman of the North?" I say.

We walk down to where the pools are and I'm just about to shout 'HAPPY BIRTHDAY' out loud so everyone around can hear when I remember: AVALANCHES! So I walk over and quietly say, "Would you like a 100% mineral salt and clay facial mask? Come on and I'll treat you."

As I take her down to the edge of the pool one half is very clear turquoise water and the other half is milky looking. Yet, when you scoop it up it is clear as glass. That's where I dig up pure mineral clay and she looks like one of the Native American Medicine Women when her face is covered.

I tell her that she has to lie in the sun, let it dry, wash it off and she will feel brand new. I also tell her that in shops back home this would have cost a fortune. This is pure, fresh clay. As it is her birthday I won't charge her. I can tell she is enjoying the fun. She looks great when she smiles. Never for one moment did we ever think she was going to spend her 59th birthday on the other side of the world at Mount Cook, New Zealand. Until last night, that is.

MAGGIE:

It really has been a birthday I will always remember. I loved every single minute of it. I toasted the day with my Run for the Hills ale and jolly good it was too. We took lots of photos. I will let Pete tell you about his dream and what we saw at the top of the mountain.

One thing I am realising is that we really can do anything we want. If we put our minds to it, use our words with tender care, be kind in our actions and follow our hearts and intuition. Believe. The Universe wants you to be all these things.

PETE:

Mount Cook is the highest peak in New Zealand and in Māori is called Aoraki (cloud in the sky). It was also a training ground for Sir Edmond Hillary for his Everest expedition. Who knows we might end up at the foot of Everest one day.

The water here is so pure you feel regenerated just drinking it. That night we have a birthday meal at the resident Hotel and a birthday drink. It's been a great day. Many Happy Returns Margaret.

When we wake up and get ready to leave, a couple and their kids have spelt the word PEACE in huge letters with loose stones lying around. I think it is great looking and I put a peace sign below it. I love peace signs. I do them everywhere I go and at any time. I've put them in every country I have ever been in: shells on the beach, twigs, leaves. You name it, I've used it. One day I'll put all the photos in one album.

Last night before we went to Mount Cook I dreamt that a wolf was walking beside me. Later, I dreamt that loads and loads of them were covering all the hills, running over them, coming towards me. Why am I telling you this? Well, I was telling Margaret all about it, when to my surprise, there up on the mountain I saw the snow seemed to have shaped itself into a wolf. It stayed there all day. I decided to call it Mountain Cloud after the mountain and the mist. Margaret quipped, "That'll be the leader of the pack then."

"The King," I replied. As we walked on Margaret points and says, "look." At the bottom of the cliff, there is a big container with the words 'YOU CAN DO ANYTHING WITH A ROYAL WOLF'. I take a photo beside it. Another weird dream.

Do you think I am an alpha wolf then? I ask Margaret. You better believe it she says. Hmmm well then since I can do anything…

MAGGIE:
The roads in this area are long and straight and very well maintained. They are a pleasure to drive on. There are not too many New Zealanders with road rage here and if we encounter a few, they are able to overtake easily. The problem occurs when the roads are narrow and windy, as they often are. Busy roads with tourists admiring scenery. I suppose it's the price they pay for having such a stunningly beautiful place to live. I suppose too, there must be such a lot of tourists and it must get frustrating. We think it is still busy just now, but, really we have been told it is quite quiet now as it is late Autumn. The weather is kind to us once again.

We are heading down to go to Milford Sound in the Fjordlands. A kind lady in one of the many charity shops we visited recommended that we visit Glenorchy. That there were many great tramps there. That's the word they use for walks.

As you would expect they have lots of different words from UK or Australia. Jandals are what we would call flip flops. Chook means chicken (chookie used sometimes in Scotland). They use the catchphrase Sweet As…to mean cool, awesome or no problem. What is really great is that there are lots and lots of Māori words in use.

Apart from visiting a Winery, our only stop is at Cromwell, a pleasant enough town. It's there, at the Information Centre, we book our tickets for a bus trip to Milford Sound. We were advised that it's the best way to do it since it is a long drive up and back down the same road and there is so much to see. The forecast is good for tomorrow which is seemingly very rare. Usually it is raining and waterfalls cascade everywhere.

We then head down towards Arrowtown for another LOTR location. It is a pretty town and we enjoy wandering around, but it is really the Fords of Bruinen we came to see.

We can't find a bridge to cross the river. Such is our determination, we find a branch and cross over like a couple of kids. I enjoy watching Pete enjoying these locations. It means so much to him as the book has been such a big inspiration to him.

Making the films were an obvious labour of love for Peter Jackson. He did the books such justice. He was the man for the job because of that love. It's one of those books if you love it you really love it. It has that effect. I must admit I couldn't get into the books when I tried years ago. I really get what it is all about now thanks to Pete.

PETE:

Our next stop is a very pleasant one as we came across a Winery we liked the look of. We sample some and enjoy looking at our road map in these relaxing surroundings. We notice that there is another Lord Of The Ring location near us at the town of Arrowtown.

Arrow River is where the Ford of Bruinen lies. That's where the Ringwraith charged in to capture Frodo and the ring by attempting to cross the Ford as they chased after him. Arwen ferried Frodo across on her Elven steed Asfaloth. The river rose and engulfed the black riders Ringwraith (remember the herd of white horses in the waves) I can't wait to go to it. I find it quite easily as it's only a couple of minutes from the van.

As I'm standing by the river taking a photo and waiting on Margaret the ring falls off my neck again. Right in front of me. This is real weird I think to myself as I deliberately tied the knots again real tight. It's getting beyond a joke. I am beginning to feel like Frodo and I am getting eaten alive by the insects which feel from the dark side. They follow me endlessly every day. I pick the ring back up, tie it once again as tight as I can. That night I try out my new lotion. As I fall asleep I can feel my lotion cooling me. I hope this is finally it.

Sounds Good to Me

MAGGIE:

Onwards we go as we have our bus trip tomorrow at 8am prompt. We drive through Queenstown and down the mountainside road towards Te Anua. Once again we need to find a campsite as it is getting dark.

We see a sign and head up just relieved to find somewhere. I can't really see a thing really as it is out in the countryside and quite poorly lit. I can see shapes that look like other campervans. I don't quite like the feeling here. It will have to do as we are getting a bus early tomorrow. We need to be up and away sharp. I am glad to lock all the doors. I haven't really felt like that before, although we always do lock them just to be on the safe side.

When the alarm went off, we could see outside the shapes turned out to be old bangers dotted about and not other campers. There were no facilities. We were glad to be leaving early.

As soon as we got on the bus we knew we had made the right choice. We could just sit back, relax and listen. The bus driver was a lovely guy and he gave us an interesting commentary as we drove past Manuka trees, huge areas of that yellow grass, misty mountains. There were clouds of mist hanging over the top of the mountains, but he assured us that these should lift as the day goes on. The weather forecast is sunshine.

We had two or three stops to see Rainforests, waterfalls, mirror mountains and lakes. My favourite stop was to see the kea. These birds are actually native to New Zealand and they are audaciously clever. It is not really encouraged elsewhere to humanise them, but here they are allowed to interact with us. I haven't laughed at a bird so much since my friendly kookaburras. They enjoy coming onto the cars, but then they refuse to back off until THEY are ready. It kind of sums up that addiction to adrenaline. Even their birds have it. We watch as a lady is forced to drive off with a Kea holding onto the roof rack at the front of her car roof thoroughly enjoying the ride.

PETE:

When we wake up the next morning, we discover that it is just a field we are in. It was dark when we arrived last night and we couldn't see a thing. Margaret decided that she would run up and pay it and when she came back down she didn't quite like the feel of the place. Well, her senses were right. We were in a field with no facilities at all and some old rusty bangers dotted here and there which we thought were other campers last night. We just high tailed it out of there.

We are heading for the bus. It's taking us a trip to Milford Sounds in the Fjordland National Park. As we're driving along we slow down to enjoy the view and there in front of us, is none other than the famous misty mountains. Our first glimpse of them. I check to see if my ring has fallen off, but it hasn't. There are four different locations for four different scenes of the misty mountains, so we wonder if we will see the other three. The mountain we're looking at is called Mt Aspiring and it stands proud in its surroundings. I keep looking back as I get on the bus. The person who told us you will need about half a dozen cameras wasn't joking.

The trip will take it around four hours to get there with two or three stops. Then a sail round the famous Milford Sound and then three hours to get back. Our first stop is a walk into a rainforest with giant ferns, weird looking trees and lots of birds. You can see why the Silver Fern is an emblem of New Zealand. It's a real braw looking thing. Some of these ferns are millions of years old, continuously self-growing and the giant fern is the size of a palm tree.

I don't know why I like the Rainforests so much. Maybe it's because I saw the Jungle Book as a young boy or A Million Years BC or something. Whatever it was didn't warn me about bites. My new lotion is getting better and better. It cools the sting and eases the itch, but it doesn't actually stop them from biting me. I will keep on trying to find an answer.

Once again as I come out of the Rainforest, I thank New Zealand for not having any poisonous snakes, toads or spiders like they have in Australia. They do have loads of annoying, nasty sandflies which can make you miserable, but at least they don't kill you.

It's back on the bus and onto the next stop before Milford Sound. At the next stop, we are told to watch out for a bird called a kea. They have the body of a hawk and the head of a parrot, only not as colourful. It's not long before we see them. They come right up. Sit on cars and play with the wipers and things. They are highly intelligent. In fact, scientific tests have proved that they are one of the most intelligent birds on the planet and they are not afraid of humans. They are just like your typical adrenaline seeking New Zealander. As one lady started her car a Kea refused to budge. She drove off quite quickly. The Kea jumped on the roof, dug it's claws in the rim of the roof, put its head down and wings out. All that was missing was a pair of goggles.

MAGGIE:

Our bus driver told us that most days it is pouring with rain in this area. He says that it is equally beautiful in its own way because there are countless waterfalls creating their own magical symphony. I can understand if you are from the outback of Australia you would love a complete day of rain with dancing waterfalls all around. I can't help feeling that the cloudless blue sky, we are lucky enough to have now, is the better option for us. We see enough rain back home.

When we arrive for the boat trip, the sun is still shining and there are stunning waterfalls. Just not to many as usual. The sea is calm which means I can enjoy my packed lunch. I always wanted to go to the Norwegian Fjords and I feel as if I am as good as there. It is little wonder this area is called Fjordland. Its certainly is place of natural wonder. When we get to the edge of the Fjords, the sea turns into a churning, mass of energy and I am relieved when we turn and head back to calm waters as my stomach is starting to churn too.

Part of our trip is to stop at a Sea Observatory to behold some of the natural inhabitants of the Fjords. They can be observed in their natural habitat. Sometimes if you are lucky, seals appear, especially if there is a shoal of fish.

On the way back down the same road, we go back through the mountain tunnel. It's an amazing feat of engineering. We also stop for a couple of viewpoints. Looking back our stunning photos at Milford Sound we cannot believe our luck with the weather.

PETE:

When we reach Milford Sound, we get our packed lunch and head out on the boat trip. Once again the weather is on our side. Not even a breeze. We get to see everything in its glory and it's like something out of The Lost World. A pterodactyl wouldn't look out of place here or a dinosaur. The unexplored, tree covered mountains surround us.

A fjord and a sound are different. A fjord has straight high cliffs on either side, made of glaciers, shaping them like the letter U, whereas a sound is made by a flowing river through a gully making it a V shape. I get told that the government introduced Elk and Moose into New Zealand years ago. I'm so pleased to hear that. I knew I saw one! He also said that expeditions used to go into the nearby rainforest looking for a fabled huge moose. Legend has it that it was as big as an elephant and had the biggest antlers ever seen. People still go into the rainforest, some of it still unexplored, to look for it. Rumour has it that you can hear it's bellows regularly, but it's never been seen yet.

Our boat stops at a little gangway and Margaret produces two tickets. We're heading into the Underwater Observatory. It's all live action and changes every second. This is where they conduct tests on the water and check the different marine life for disease. We see some odd looking fish. We also see a huge shoal of mackerel. It's a mass of bodies all moving as one. Similar to seeing a flock of starlings in the air. What a great day out.

We feel real satisfied with our day. If that's not enough, according to our map, tomorrow we're heading to Fangorn Forest, home of old Treebeard himself. On the road, we stop and take photos. We just can't help it.

We come across a small village called Manapouri which has a campsite with a terrific setting and very well kept hedges. It must be a full time job looking after them, they are so well shaped. As we go into the communal lounge we soon discover the wifie runs a tight ship. Her place is spotless, she knows every blade of grass and every little thing going on. All in all: a good clean, firm but fair, campsite.

MAGGIE:

Once we are back at Te Anau it's still light. We go for another walk as we really like this town. We come across a park with a bird sanctuary where we get to see some Takahe. Seemingly across the water in the uninhabited mountain areas are the last Takahe living in the wild. The stoats haven't managed to cross the water and so their eggs are safe.

We are heading down to Manipouri a nearby village. That is as far south as we are going. It's time to start heading up towards the west coast. The first thing we come across at Manapouri is a large tin sign. It's a giant possum. Even if it made of corrugated iron, it still has those big, sad eyes.

As we explore we come across another of our Royal Wolf sign howling at the moon. Afterwards we head to find a campsite. It makes a change to be looking while it is still light. The one we find has amazing manicured hedges. Can't help thinking about our friends who have a lovely six acre garden back home in Ayrshire. It's full of wonderful trees and hedges. Most of them are used as wind breaks which I think is a great idea. It could make a big difference back home. Mind you it must have taken years and years to grow these hedges from poplar trees, fir trees, pampas or elephant grass to name but a few.

The lady running the campsite runs a tight ship. The place is immaculate. The facilities are great, but this comes with having the wee lady shadowing you. In the morning, I was brushing my teeth and nearly jumped out of my skin when I turned round. There she was, behind me, making sure I had wiped the shower down.

Time to move as we had to be out by 10am <u>sharp</u>.

PETE:

We drive on and see something in the fields. They're Alpacas and Margaret fancies going in to see them. At first, it seems nobody is there. Then a woman came out and said it's Easter Sunday. It's an official holiday. Everywhere is closed, but I'll show you round. She gives us a tour round the fields and pens, tells us all about the Alpacas and their wool. They have to be a certain age before they can be sheared for their wool. The breeding has to be right. They are trying for a chocolate coloured baby one. They are a bit odd looking and have faces you can't stop looking at. We have a good wee chat and we happen to mention that we camped at the neighbouring campsite the night before and that they were very pernickety. She seemed embarrassed and smiled knowingly.

I notice that she sells Paua Paua shells. I tell her I have some. She shows me one that has been cleaned and polished and another that has not been touched. The difference is incredible. No wonder they make jewellery out of them. I decide that when I get time I will polish mine.

As we are about to leave the village we see a sign saying the Little Book shop. In we go. It's got the kind of old books I like. I buy one about a boy kidnapped by Red Indians. He ends up not wanting to leave them as their way of life of life is much better than us whiteys. As soon as Margaret tells her that she used to be a Librarian she takes a shine to us. She even goes to her house and comes back with a herb book for me as a present. I look around to buy another book and I spy the Official Lord Of The Rings Guidebook. Very apt. That'll do for me. It was the first of three very different Lord Of The Rings guidebooks I was to get. The reading of all of them was all the better when you actually have been to and seen the places mentioned.

Margaret gets her photo taken with the wee book shop woman and toot toot, honk honk we head onwards and upwards.

MAGGIE:

We stop to see some Alpacas and a chat with the owner who gives us a tour. She tells Pete that the cast of Lord Of The Rings had to be helicoptered to the misty mountain which we can see from her garden. She said the films brought a lot of work at the time and a lot more tourism to New Zealand afterwards. The alpacas are so cute and their fleece so warm. Nice to meet you.

Then I met, for me, the loveliest wee woman in the South Island. She had a tiny wee bookshop which caught our eye as we were heading away out of the village. She used to own a bigger book shop round the corner. When she retired, she missed it so much her husband suggested she do something about it. Hence, the tiny wee bookshop.

Here we were: two women who have loved books all our lives. We had a long chat. I told her about my library days and how I knew all the books and where they were and who liked what. How I came alive at my work. I believe in libraries so much. I actually met Pete in the library I worked and we were good friends for many years. We seemed to be reading the same sort of spiritual, travel and history books. Asking the same sorts of questions. Eighteen years later, here we are, married, travelling the world. We both did not see that coming for one minute.

She got what I meant about the books. She said she took pride in the fact she seems to know what book a person might be searching for, even if they didn't quite know themselves until they read it. That's what the true love of something does I say. It creates a kind of magic.

She goes and gets a wee herb book for Pete and looks knowingly at him as if there is some nugget or treasure inside which he needs to find. She gives me an Edward Rutherfurd book called Sarum with many pages, which I haven't got round to reading yet as it is packed away somewhere in a box at the moment. I know, for sure, I will read that book at exactly the right time and get exactly what I need from it.

I could have stayed and talked for hours, but it's time to move on. We say our fond farewells, all the better for having met each other.

PETE:

We drive a good bit in the open country with hardly any traffic. According to our map, Fangorn Forest is not that far away now. It's pretty remote. We finally reach it and all the trees look as if they've got faces. It's Fangorn alright. I decide to myself which one is Treebeard and politely ask him if we can tie a clothes line onto him to dry our clothes. This is great. Wait till we tell them back home. We're hanging our washing out at Fangorn Forest. We're very remote, so it's not as if we will be annoying anyone.

All of a sudden a guy comes round the corner in a pickup truck. He stops in front of our washing. He says he manages this land. I thought I saw someone drive up this way. We're travelling around I tell him and couldn't resist coming to see Fangorn Forest. When I first came here, I took over the management and never knew anything about Lord of the Rings so when people asked me if it was Fangorn Forest I'd say no, you'll need to look elsewhere. However, I know all about it now. In fact, jump in my pickup and I'll take you to the spot where Treebeard walks out of the forest. Just ignore the dogs.

He drove a fair distance down a track. Finally, he got out and said there you go. That's where all the big filming took place. Tell you what, he goes on. You can bring your van and camp the night if you like. Nobody will see you here.

Eh don't know about that I think to myself. A bit too much of the Dodgy O'Doyle for my liking.

I'll give you a lift back to the van and you can think about it. We're in the van and I am thinking to myself this is a bit of out the way. What if he's a nut and thinks we are loaded or something. Then he says if you do stay lock the gate behind you as you go in, just to stop anyone from coming down.

That's it I think there's no way we're staying. Then his mobile rings and he says hello and a female voice answers on loudspeaker. Then he stops speaking in English and starts speaking in a foreign language. When we stop, he says it's my daughter. We always speak in Hungarian when we can because my wife is Hungarian.

Well, so do we talk the same language. It's called telepathy. As soon as he goes, we pack up, but not before I say thanks to Treebeard for looking after our washing. I'm careful to thank him in English and not Entish. As that would take days. No offence to anyone, but we are in a bit of a hurry and with that we leave Fangorn Forest.

MAGGIE:

We doing a loop back up to Queenstown and then take a left up a one way road to Glenorchy. We hoping to find the location for Fangorn Forest on the way. Whatever road we take, there are always mountains and they are definitely misty looking this morning. The sat nav points up a track road which goes on for miles. We do eventually find Fangorn Forest and get to meet the Tree-Ents.

On our way to Queenstown, we notice a rainbow. The beginning seems to be coming from where we are leaving and it ends up in the misty mountains to our right, where we are heading. Perfect. Follow the yellow brick road.

Our journey is like a big horseshoe as we go up one side of Lake Wakatipu cross over to Queenstown and then head up the other side of Lake Wakatipu up to Glenorchy. This is where we were told there are lots of great tramps.

Och Aye the Noo

PETE:

We get proof that this scenery is straight out of Lords of the Rings. As we are driving up the road to Glenorchy, we see a Tour Minibus for Lords of the Rings parked in a lay-by. Folk have their binoculars out. The road is too busy for us to stop, but I am happy in the knowledge that there are more opportunities further ahead to see Isengard, Lothlorien and Amon Hen.

Glenorchy sounds Scottish so we go for a wander round the place and we are not surprised to see street signs with Scottish names such as Glasgow, Ayr, Perth and much much more. No tattiebogle though. Talking about tatties we decide we are hungry and go to a small shop. We had forgotten to get supplies earlier so we don't have much to eat and Margaret fancies an omelette. She picks up half a dozen eggs. They guy says that will be $7.50 please without batting an eyelid. We decide not to buy anything else. At those prices, you expect the chicken with the eggs. It doesn't stop us from enjoying the walk roundabout though.

Guarana Rave

MAGGIE:

After a stop at Glenorchy, we drive on to find Lake Sylvan campsite. On the way, there we should come across Isengard. There are lots of opportunities to tour Lord Of The Rings locations by helicopter and tour bus. We much prefer and love the freedom of Noah.

Once we have settled we are in our beds just before 8 pm. It is dark by then and no campfires are allowed at most areas. There have been no droughts like Australia. It has been a dry summer though and fires are always a risk in these conditions.

Pete looked out and saw a huge lorry arriving across from us. We didn't expect to see that way up here. When a bassline began to thump, we just looked at each other puzzled. Then the penny dropped. It was the Easter Weekend and there was a rave going on. I had never been to a rave and had no intention of ever going to one. That night we didn't really have a choice, unless we got up and drove on. I looked out the earplugs I had given me for my travels. It is the first time I have used them and I am pleased to report they worked.

In the morning, once the ravers had left, we go for a long walk to find Lake Sylvan and see what birds we can see. We mostly see possum traps. We do see a lorikeet of some description. We also some wee fantails or it might have been the same one following us. They love to show off their tails. Sometimes you think they are actually going to land on your hand they come so close.

We have a pleasant time tramping. One thing about New Zealand is that you can go off the beaten track as there are no snakes or dangerous creatures. There are however quite often signs ordering keep to the track. This is due to the fragility of the ecosystem especially to trees and flora due to the introduction of non-native species.

PETE:

After our wee adventure at Fangorn, it's time to go in search of Isengard. We head to a campsite at Lake Sylvan. It is nice and quiet out here in the middle of nowhere. There are a couple of other vans. We make our omelette and then it's time for bed. Everyone in Australia and New Zealand seem to go to bed early and so we have adopted the same lifestyle.

I hear a noise outside and I look out to see a huge lorry manoeuvring round the back. That's some size of lorry I remark. It was about thirty to forty feet long. We could see the words Guarana Energy written on the side. The lights are switched on inside it to reveal canvas sides which we could see through. People were walking about inside and outside the lorry. Guarana is a plant from Brazil I tell Margaret. I know this because of my herbal medicine studies. They use it to give energy. As soon as I say this a loud blast of music comes on. It was a rave and a loud one at that. Very loud. Sounds like the records stuck and people are jumping around inside.

At first, we thought its Easter Weekend, the young ones need to do something and not just us. At 4 and 5 in the morning, it was still in full flow. It actually went on for twelve hours starting at 8pm and finishing at 8am. They must have drank a ton of Guarana.

Later on we take a walk to Lake Sylvan. I rub my insect lotion on my body, legs, arms and ears. It seems to be working as I am feeling not too bad, but feel I am still looking for that vital ingredient.

Paradise

MAGGIE:

It's time to find Paradise, although I must say that I feel I have been in Paradise for the past couple of weeks. The scenery is just relentless in its beauty and I never tire seeing it. So I am thinking I don't see how Paradise can outdo what I have seen every day since arriving in New Zealand. At Mt John Space Observatory, it was paradise to me. I actually brought me to tears. Tears of happiness. Mount Cook was paradise to me. Milford Sound was paradise to me. I am still not immune to the beauty everywhere. I have often stood with my mouth open, speechless.

So off we go in search of Paradise. We have to drive down a long track to find it. We have to park in a DOC campsite with a dunny to find it. We have to cross a Ford to find it. We have to go beyond the signpost saying Paradise to find it, but when we do? Yeah, of course, it was worth the effort.

There are the magnificent mountains we have come to expect but never take for granted. There are lush green trees. There is even a white horse which, with a swish of its tail, could be Gandalf's horse, Shadowfax. The magic is everywhere and Pete is loving it.

PETE:

I am really excited about seeing Lothlorien, not only because I want to say I've been to Lothlorien, but to get there you have to drive to a place called Paradise. Again, it's an off the track kind of place to get to and wonder if we are on the right dirt track because we have been driving on it for a while. A van is coming towards us so we wave it down and sure enough Paradise is just round the corner.

We stop at the little campsite and we can't see any sign of Paradise. The snow covered mountains are stunning. The lake is large with lots of ducks and swans. The surrounding trees look like they are suffering from some kind of disease or from a fire or hurricane or something. They look like skeletons. There are no leaves and they are covered in lichen.

We decide to follow the dried out riverbed to see if it will lead us to Paradise. So we put our boots on and FOLLOW DRY RIVERBED. We eventually come out at a Ford, not too deep, so we cross it.

Eventually, we come across a young Asian couple taking photos. There is a signpost which actually says Welcome to Paradise. We are actually in Paradise and it lives up to its name. It's not long until we are taking photos as well.

At the top of the hill as the sun filters light through the leaves, we can see Isengard from the other side. We will go a walk in the morning to explore further.

We went further up the Paradise Road today. There are horse treks available, there is even a nice, big white horse, if I did get tempted. There are treks through the woods, but I would much rather walk. I see the Lothlorien wood and it is easy to picture the beauty and transport yourself into a magical world for a while wherever you are. This is what we must do for it is time to leave.

MAGGIE:

So we say farewell and head back down the road via Glenorchy and Lake Wakatiku and eventually end up looking for a campsite as we feel that we have done enough driving. The Youth Hostel is not for us as the showers look, well dodgy.

We go to Lake Hotapa campsite for a bit of TLC. We got talking to a couple at a big outdoor campfire. They are New Zealanders. She talks about the busy roads down here and I can feel the anger coming off the woman. Then she says it's only mountains like she can't understand what all the fuss is about. I suppose if you are born into it and it's on your doorstep all your life you could take it for granted. I tell her I have seen many lovely places such as Switzerland, Iceland, Norway and the Scottish Highlands, but I have never seen so much beautiful scenery, day after day, as here. She looks surprised.

Don't get me wrong I have many other places I would love to see. Canada is one of them and it so happens a young Canadian couple come and sit at the fire. We really enjoy swapping stories about our homelands.

PETE:

We travel back, through Glenorchy and I have a beer as it is another beautiful day. Then back down the same winding road towards Queenstown. We feel that's enough for one day and look for a campsite.

We end up at Lake Hotapa campsite. A nice place with the usual showers, kitchen etc. However, we notice quite a big fire on the go, at the lakeside. Since we love a campfire, we head down, where we discover that the couple there were hogging it as if they owned it. We spoke to them all the same. As time went by they left and another couple came by. They were all the way from Canada. The guy was working and living in New Zealand for the year in some sort of student exchange thing and his girlfriend was visiting him. They were a nice couple, much friendlier than the two old farts that had just left. The girl went on to tell us her grandparents were Scottish/Irish and every Sunday they would argue over what country brought over the tatties. LOL.

We drive on back through Queenstown and then up towards the west coast. The road climbed higher and higher, going round and round and up rather than down. I was so glad it wasn't me driving. Full marks to Margaret on that one.

As per usual we need to find a place to park as it is getting late and don't forget it needs a toilet. We surely must be near the top by now. We turn a corner. It's beyond belief. I can see the huge full moon, only it's below us. What the…? There's an airplane below us too with all its passenger lights on. We are so high up. Looking down at them reminds me of the Fairground Chairoplanes. It was a bit surreal.

MAGGIE:

We land at Queenstown around noon and decide to do the laundry. By the time we get a parking place etc. etc., guess what? Yes, darkness is falling and we are heading towards the Haast Pass up the west coast. It's a quite a distance so we know we won't make it tonight. Still we want to find a camping ground further up the road.

When we said up the road, we didn't mean literally, but that's what happens. When we take to the road, it is dark. We find ourselves going round and round and round for what seems like an eternity. As we turn the corner we see the full moon in all its glory. I love the moon, so I am delighted that we are actually above it. That is definitely a first for me. To be above the moon.

I think of the poster at Manapouri of the Royal Wolf howling at the full moon. I think of Mountain Cloud, Pete's guardian alpha wolf from his dream. You can do anything with a Royal wolf. I can do anything I say to myself. If I put my mind to it, with all of my heart, step by step. Keep the faith babe. Keep the Faith.

That had to be the highest campsite we have ever camped at so far. Right at the top we found a free DOC campsite. It was very busy. I was careful not to park too close to the edge, as I have a feeling we will have quite a view in the morning. Sure enough, we are extremely high up with a view right over the valley. It's not the kind of campsite you want to hang about at. We must say we are extremely grateful to the New Zealand Government for providing these places free of charge. I wish our Government would do the same.

PETE:

The Haast pass is our next destination when we will eventually get to see Mount Cook from the other side at the Franz/Joseph Glacier. We have a couple of stops on the way the second being a visit to the Mirror Pools.

As the name suggests they are in a constant state of mirroring everything around them since the water is so clear. Honestly you could comb your hair and shave in their reflections it's like a photo sitting on top of another photo. I wish nobody was looking so I could try standing on my head to see what it looks like, but as we're on a swinging bridge, we don't bother.

On the way back down, we discover there is another Rainforest walk to where our van is parked. We decide to venture in. It looks very dense in places and I am wishing I had put my lotion on. I put my socks over my jeans, keep my hands in my pockets. I don't plan to get bitten today thanks. In we go, and it is everything I expect. I have been in a few now and just love the thought of all these undiscovered flowers, herbs, trees and even beasties to be found. I go off the wooden walkway and do a bit of exploring myself, looking at the moss covered trees and stones that, for all I know, could be ancient steps. One of my things on my bucket list is to visit Machu Picchu and for some reason this reminds me of it with its dense forest and giant ferns.

I eventually get back on track. We take the path out of the forest and head on our way, but not before we have a picnic lunch. We drive for a bit and as usual in New Zealand it gets dark pretty quick and we are left looking for a campsite. At least, we know there are plenty and we find a small DOC campsite with an honesty box and settle down for a sleep.

OUCH! that was quite sore. Something just bit me right in the middle of my spine. It feels like a needle of some sort. Sharp and deep. OUCH! I pull myself together and think I better look my lotion out in the morning for that one. With that I fall asleep. When I wake up, I don't feel too good. Sort of sluggish. These bites are knocking the stuffing out of my immune system.

MAGGIE:

We drive on to Wanaka which is next to Lake Wanaka and is very picturesque. In fact, we spend a couple of hours there. There is a huge sculpture of a hand which we are allowed to sit in. So we do.

It's on to the Mirror Pools which we have been recommended to visit.

I think it is our friend, the Lonely Planet, who suggested we do. The roadside car park is full which nearly puts us off going. As we turn the corner there is an overspill car park and in we go. There is an option to walk through a path in the woods and then it's another fair walk through another wood to get to the pools. It is busy since it is the Easter School Holidays.

It's another lovely day weather wise which allows us to see the pools in perfect conditions. They don't let us down. The water looks jade green from the bridge, but it's very, very clear. Luckily the swing bridge is only about a hundred feet long and not like that other ridiculous one we visited at Murchison. There is a great deal of etiquette involved in crossing a swing bridge. How many at a time, who goes first etc., otherwise you could be waiting forever on a busy day like today.

After having lunch at a picnic table next to Noah, we drive on. We feel so lucky to be living the dream.

PETE:

We come to a town called Fox's Glacier township and straight away I remember an advert back home for Fox's Glacier Mints. So that's where the name came from. I always wondered what a fox had to do with a polar bear and mints. The town has glaciers around it, but they are melting. Although it makes for great photos, it's alarming the rate that they are melting at. Through a little gap you can see another angle of Mount Cook which means, of course, another angle of the misty mountain. The scene where good old Gandalf leads Frodo and company through the snow and they end up having to enter the caves of Moria comes to mind. Man, I just love this: actually seeing all this Lords of the Rings stuff.

Now true to form it seems like I've been bitten because my spine's sore and twitchy. We head to a pub and discover its name is Snakebite. Very apt I say to Margaret, thinking about my bite (it wasn't a snake, but it sure feels like one). We look at the snakebite menu: a snakebite is cider and lager. We used to drink it back home, but they didn't have names like these ones. You could have a VIPER, COBRA or even a TAIPAN.

Well now, remember the wee Rainforest woman in Australia? She got bitten by a brown snake AKA a Taipan and was talking in tongues. It seems the tongues she was talking in were Māori. Be Safe, Be Strong and God blessings. We just had to have one and toast her (although I'm still a bit dubious about her living in Bethlehem and Judea) Cheers.

MAGGIE:

We heard a lot of scrambling about on our roof during the night. Turns out we think it was a Kea. Pete saw them in the dark when he got up for the toilet. It might have been a Kaka which are cousins of the Kea. They are a bit rarer so we think it was the Kea. We feel so happy to have seen them in the wild.

We would love to see a Kiwi in its natural habitat. They are nocturnal. Like our friend the platypus they are so rare and shy it is notoriously difficult to see them. We are going to give it a try the first time we get the chance. We have heard that there are more of them up on the North Island, but it's the same old story about the stoats. They are on the endangered list.

We get to the Franz Josef township. It's not that big. I imagine it is very busy at high season. Not so many people have ventured this far up at this time of the year. It is definitely cooler up here. There is a chance to see a kiwi in captivity including some younger ones. We decide to wait and see how we get on up the road.

We plump to go for a snakebite as there is a bar in town called Snakebite which specialises in them. Again recommended by our friend, Lonely Planet. The food is pretty good too and I have my first pizza in a while. Snakebite used to be popular back in the day when we were young. I am not that much of a drinker. I never really bothered that much with alcohol in my younger days but I am kind of making up for it with my love of real ale. I want to try one of these as a homage to our wee prophet Rainforest Healer Woman. We order one each. We toast her and thank her for the Karakia protection she gave us.

Killing Paradise

PETE:

We end up at a place called Pukekura. It's an ideal name because that is what I feel like doing. Puking. I don't know what's going on, but I haven't felt right since feeling that needle thing in my spine. I am quite weak and to be honest not very hungry either, but I could drink an ocean dry. I am getting itchier and itchier, but it's a needles and pins kind of thing. I feel it all over my body, even my head.

As per usual in New Zealand it was completely dark when we arrived last night. I head into the cabin for a drink of water and see a poster. Welcome to Pukekura the smallest town in New Zealand. Population 6. The water is safe to drink. Well thank God for that because I have just swallowed about a gallon.

Next morning, we are taking the opportunity to clean and air the van. Just at that a guy pulls up and asks if we are OK. I think it's because he noticed everything is out of the van. I explain what we're doing and get talking to him. He is a trapper. He traps possums for their fur. He traps them because there are millions of them. He goes into the woods at night with just a knife, a torch and a pair of jandals (sandals). He never buys meat. He hunts it. He and his wife and children grow their own vegetables and use rainwater where possible.

Turns out he is very angry with the New Zealand government as most of the farmers are. The government spray a poison called 1080. They use helicopters (we saw it in Australia as well). It's killing everything and poisoning the rivers. Look it's like this he says: 1080 kills insects and possums. Birds eat insects, large birds of prey eat possums, dogs eats possums and so forth. It also in our water and that's why we only drink our own filtered water. He goes on to say that he and some friends held meetings and protests. The government, he says, has deaf ears. They don't want to hear about it. They advertise New Zealand as paradise but they are killing it from the inside, where you can't see it. Slowly. Buy the DVD called Protecting Paradise by the Graf Brothers he says and off he goes.

I get another drink and fill up our big bottle. I'm still thirsty.

MAGGIE:

I heard a story which I didn't want to believe. The New Zealand government are dropping a poison called 1080 all over the south and north islands using helicopters. They believe it is the solution to the possum problem.

We had noticed billboard type posters here and there on our travels with BAN 1080 on them. We didn't know what 1080 stood for or indeed was, until we got talking to a young guy at Pukehura.

He is a trapper who hunts possums for their fur. He talks about the poison they are dropping and also putting in the wooden traps we have noticed in the woods. If what he is telling us is really happening, then it will be in the food chain from plant and insect up and in the beautiful, pure, turquoise, glacier water. There are always two sides to a story. In fact, it depends on how many are telling the story how many sides there are. I don't want to believe that this is happening.

We move on up towards Hokitika.

Haka and the Little People

PETE:

We leave the smallest village and the population goes back to six. We head up to Hokitika. It's where most of the green jade is found. Maoris make necklaces and Koru from their precious greenstone or Punamau as they call it.

In the morning, Margaret always asks me if I have any dreams and I told her about the dream I had last night. Our campervan is called Gabie after a friend who passed away. We were going to call it Gabriel after the Archangel with the intention that he protects us. When we found out that my friend Gabie had been going to buy the actual van the week before he died, we shortened it to Gabie. Last night I dreamt that Gabie was digging a garden with a small spade. In my dream, I asked him what he was doing as all the digging looked the same shape. This is your design, he said. When I woke up, I tell Margaret it looked like a figure 8, but not quite or maybe it looked like two S's (esses) touching, but not quite joined.

As we walked around this cute place called Hokitika we came across a little shop that sells jade. You can go into their workshop and create your own design. They'll let you do it or they will do it for you. I remember that I have the paua paua shells and wonder if they can show me how to clean the calcium off the shells.

We walk in and these two guys are huge. One is twice my size and the other is twice his size. They are the happiest looking New Zealanders we have met. All smiles and jokes. A great vibe. Is this all your own stuff we ask. Yep we're cousins and this is our own business. I found some paua paua shells can you help me to polish them and tell me how much it will be? Grinning, the smallest of the two, jumped up took a shell and said. Come outside and I'll show you the old way for nothing.

He started to rub the outside of the shell vigorously against the concrete step. Use any stone and you'll save a fortune he said. The biggest of the two said we used to gather these as kids. All the time we were playing and polishing them for fun, mum would say I'll have that. She was probably selling them. So now we polish them ourselves and make groovy looking things from them and jade for a living.

The two of these giants were laughing and grinning like school kids. I just love them. Is it OK to ask for a photo with you guys I ask. Sure one of them jumps up, opens the glass cabinet and takes out a huge piece of jade the size of a small chair leg and throws the other one a small version to which he says hey I want the big one. The two of them laugh then all of a sudden they let out a roar with eyes bulging and tongues sticking out. I'm getting a real life HAKA from the Maoris. I'm stunned.

Folk pay for this and here I am getting the real deal from both of them. We all start laughing and get a couple more photos. It's been a real pleasure and privilege talking to these good, kind people and as we are leaving we hear the small, big one say to the big, big one, in the good natured way they had 'that was the little people'.

MAGGIE:
Hokitika is my favourite place on the west coast and not just because I love its name. Apart from at Kaikoura we have not really met many Māori people.

We find out that only 4% of the Māori population live on the south island. I remember the Canadian canoe sculptor telling us that the white settlers sold or even gave guns to a tribal chief from the North Island, encouraging him to invade the south whereby he proceeded to invade the south and wipe out the tribes living there. Perhaps this is part of the reason the percentage is so low. I wonder if the person who invented the gun realised at the time just what horrendous slaughters they would be used for in the future? Time and time again.

We met a couple of Māori guys who were big in stature and big in heart. The real genuine kind of people it is a privilege to meet. The ones you never forget like Priscilla from Gladstone. They were so full of joy it was contagious. We hadn't laughed so much since Australia.

They had their own business doing the thing they loved. Now that is what life is all about I thought. They made us so welcome without once trying to sell us anything. Then they made our day by doing a HAKA for us without us even asking. As we left we heard them, good naturedly, calling us the little people which made us laugh as much as them. We must have looked like a couple of Hobbits or Leprechauns. So many folk on our trip, on hearing our accents, have thought we might be Irish. They were such friendly giants.

We then headed to a local museum. An English guy got chatting to us. He asked us whereabouts we came from as, I hope you don't mind me saying, your accent is quite coarse. I felt like I was back home for a moment because over here we have been told over and over that they love our accents. It kind of jolts us back to the insidious bullying we have been subjected to by the 'English'. He and I are of a certain age and I think that may be a factor. Don't get me wrong I have loads of lovely friends who are English but, as a nation, the south of England really have believed that their way is the only way and any other way is inferior. I think the tables have turned now, but the damage in Scotland, especially the lowlands, has been done. Anyway, I think he realised what he had said and changed the word coarse to strong. Aye, we definitely have a strong accent.

The Old Mines and Shelob

PETE:

We are heading up the road to Picton via Reefton. Driving along we see a sign which says Taylorville. As Taylor is the name of one of my grandsons. We stop to take a photo. Just along from here we had seen a signpost Waiuta Old Mining Village. I'm an ex miner. I worked in the mines for ten years. I want to see this place as in my young days I was brought up in the Glenburn Rows, Prestwick. Three rows of houses with a Steamie (laundry room) in the middle one. It was built specifically for miners. I'm glad Margaret is driving as its way into the forests, high up a narrow, steep, winding road. There are huge ferns everywhere. It was windy through the night so there are broken branches hanging over the road. This road is rarely used.

We're glad when we finally get to a clearing. There in front of us, is what can only be described as, a bunch of derelict buildings. I thought about the road we had just come up and wondered how on earth the miners managed to get their materials, food and everything else up. This place is intriguing me though. How on earth did they manage to live here? As we walk around there are more derelict buildings. One of them says Waiuta Cottage. I brush away hanging leaves and see another sign that says. This Cottage is Occupied—Please Keep Out. I go further and another sign says Area Closed Due to Unsafe Mineshafts and Tunnels.

The information Board has photos of Waiuta in 1923. It's a picture of the whole village. It's astonishing. There was a football park, swimming pool, post office. They even had a club to go to for a drink. There is a network of roads connecting and there is a school at the far end, This was a town in the middle of the mountains. They must have travelled miles to get here. The notice says at one time there were thousands here, cut off by the weather and geography until it was time to head back with their gold. They would cash it in, get provisions and head back again. Today there's hardly anything, except these few reminders of a hard life. I enjoyed that. Maybe it was because I'm an ex-miner, as was most of my family.

MAGGIE:
We don't have that much time before we sail for the North Island on 30th April. We have heard that the coastal road from Westport upwards is a beautiful drive and there are lots of lovely tramps. We just don't have the time left to do it justice, so we decide, if we come back to New Zealand, we will do it then. We are heading towards Reefton an old gold rush town which has managed to survive the fact that the gold ran out a while back.

Pete is not feeling so well. He is not complaining, but he has been bitten by sandflies nearly every single day in New Zealand. I am relieved to say that they hardly ever come near me. It is starting to take its toll on Pete. He looks very tired and I feel he is summoning up the willpower to enjoy this day as best he can. He doesn't believe in projecting his discomfort or misery to others, so he keeps his feelings to himself.

PETE:

I had to put my insect lotion on, not necessarily because of where we were, but because this morning, once again, I woke up having had a terrible sleep, my head thumping, my mouth dry and my full body pulsing from head to toe. I feel like all my skin is going boom, boom—boom boom. The top of my ears were itchy. Margaret looked at my back again and this time she exclaimed. Pete! You need a hospital. I froze. Never in all my sixty four years had I ever needed to go to hospital and I don't intend to now. Margaret is adamant. She wants me to at least get some antihistamines. This feels as if it is actually right under my skin.

Walking around the old ghost town took my mind off things, but it was marred by this horrible feeling. It was as if I couldn't take anything in. Sometimes everything would look clear and other times everything seemed muddled. Man I am so dry. Even my skin is drying up.

We stop at a town called Alice (only joking)—it's called Reefton. It's got a hospital. Margaret says she's not going in, but she is not letting me get away with sitting in the van. I reluctantly go into the A and E Department to find that nobody is there. It's Sunday. Good, I say to myself, I didn't want to go in anyway.

As Margaret is looking over, I go back in sheepishly and shout quietly Hello, Hello? Just at that a guy opens a swing door and asks if he can help. I ask if he is a doctor and the answer is no it's Sunday. I told him that I had been bitten by a flea or something and can he give me something for it. I told him what Margaret said when she saw my back this morning and he says let's have a look then.

When he lifts up my t-shirt, he lets out a kind of gasp as he says that's no flea—that's a white-tailed spider that's done that.

Eh a what! A spider! Are you sure? I thought there were no poisonous creatures in New Zealand.

There ain't except for this one. It's got a white stripe down its back. That's why it's called a white tail.

Can you help?

Sorry as I say it's Sunday the pharmacy is closed and there are no doctors available. I can give you advice but can't prescribe anything.

Well thanks anyway. At least I know what it is and walk out.

Suddenly I see myself in the window and the reflection that looks back. My face has started to swell up, my nose, cheeks. My ears are worse they are ringing constantly. Is that me? JUST GREAT, NOT ONLY DOES MY RING KEEP FALLING OFF, NOW I'VE BEEN BITTEN BY MY VERY OWN SHELOB.

Walking over to the van, I remember for some reason, the wee woman in the Rainforest who spoke in tongues. She said I would be going into the heart of the enemy, but not to worry as Jesus had gone before me and I was just to keep the faith. I kept the faith alright. I don't want die here in New Zealand.

Margaret took me to the supermarket and we managed to get some strong antihistamines. The woman there suffered from hay fever and knew all about antihistamines. She had also heard of the spider and yes it was very rare. Where did you go to get that? She asks. They're rarely out of the undergrowth and tend to frequent rainforests.

Rainforests. God I love them and have been in a few. We sit down and piece it together. It must have happened at the Mirror Pools where I went off the walkway. It must have got onto my clothes and waiting till night time, stuck it's stinger in my spine. Now, a few days later it's venom is in my blood stream. I feel like death.

MAGGIE:

As we are walking through a ghost town from the gold rush days I can see Pete is making an effort, although I know as an ex-miner he is finding it interesting. His face and lips are beginning to swell. When we got back to the van, I asked to look at his back. When I saw all the angry looking red welts all over his body, I said enough is enough. You need to go to a hospital. He tried to resist. I did not take no for an answer and we headed back down the track towards Reefton. When we eventually got back on the main road and I get a signal, I am relieved to see there is a hospital in Reefton.

After we find it, Pete came out to tell me that because it is Sunday no doctors are there. He then goes on to tell me that he has been bitten by a white tailed spider. All I can think is thank God he didn't get bitten in Australia because if it was one of those snakes, scorpions or spiders, twenty minutes is often all you get to find an antidote.

Herbs and Their Healing Powers

PETE:

The next couple of days go past in a bit of a blur to me. I manage to persuade Margaret that I will fight this off by myself. I am stubborn and I have never been a patient in a hospital in my life.

It's our last day on the South Island. It really has been awesome despite all the bites and I can't wait to see what the North Island brings. We've been told that it's a bit colder and wetter. I am glad to hear there are a lot more Maoris and according to our trusty map a few more Lord Of The Ring locations zigzagged across the island again.

Before the ferry crossing, we stop here and there at some places with some great names like Ngakuta Bay, Whenuanui Bay and Waikawa. At Picton, I have promised Margaret to go to a Pharmacy. Once there I do as I am told and the helpful Pharmacist gives me some strong antihistamines.

After the ferry sailing, we decide not to bother going to Wellington. Too touristy we decide. It will probably be lots of shops. We want to see and experience the real New Zealand: exploring Rainforests and getting bitten by rare spiders.

I can't think straight, I keep hearing voices in my head 'echoes'. I keep thinking that Margaret's talking to me, but every time I look over she's driving. Fuck. I hope I am not going to die. This spider thing is beginning to frighten me. I remember saying to myself tell Margaret you love her and to tell everyone else that from me as well. We stop at a place that I don't remember much about except that I buy some Golden Rod and Echinacea in a Pharmacy.

We see a sign for a campsite at a place called Kaitoke Regional Park. As we drive in I can't believe my eyes—there is a sign with an arrow pointing to a place down the road. Above the arrow is the word RIVENDELL.

I felt stunned to begin with. I certainly wasn't expecting to see this. Sure enough Rivendell is just down the road—Every Lord Of The Ring fan's dream and of course, a place of healing. Yeah I thought in my fuddled head, even camping here makes me feel better. That night I say to myself, if I get better and live, then I will try and be more careful with my life. I don't know what's in front of me. Being careful and a decent bloke is all you can ask for, I think.

MAGGIE:

Like Pete, I am a great believer in complimentary herbal medicine. It changed my life around the age of thirty. It was then I found out about echinacea and reflexology. From the age of eight till thirty, I had just about everything you can think of from scarlet fever, pneumonia, bronchitis, measles, German measles, mumps, glandular fever. I also seemed to take every cold and flu going around. After finding out about and taking echinacea, I turned my health around. Those years of illness taught me to never ever take my health for granted. It is precious. If I can avoid taking prescribed medication, then I will, unless I feel there is no other alternative.

As we sail out from Picton, the first hour we are surrounded by land and therefore the water is calm and scenery spectacular like Milford Sound. Once we reach the open sea it is a different story. Luckily, I had taken my homeopathic seasickness tablets and we sat in the TV lounge to take my mind off the choppy sea. Pete doesn't get seasick like me. His feet are firmly planted on the ground.

Once off the ferry, I just keep driving on up away from Wellington and the busy roads. We find ourselves in a place called Upper Hut and go into the pharmacy where Pete was guided to some echinacea and golden rod tincture. He is still struggling to shake this off.

We decide to drive up the road. We didn't have a signal and I realised as per usual time was running out to find a campsite before darkness. About eight miles up the road we are relieved to see a sign for Kaitoke Reserve campsite and in we go. Then we saw the sign RIVENDELL. We just looked at each other. Speechless. We were definitely led here.

PETE:

I take a real good concoction of Golden Seal, Echinacea, Manuka honey, paracetamol and antihistamines and then lie down thinking am I dreaming or am I actually camped outside Rivendell?

My head's a mess. All night I feel restless, itchy, not itchy; itchy, not itchy; breathless not breathless; twitchy not twitchy. It's a horrible night. I imagine I'm in a campervan camped outside Rivendell. I think it's raining, but when I wake up I'm soaked to the skin, the bed is soaking and it smells horrible. I rustle myself up. To my surprise and delight I feel my head is actually clearer. I'm not so itchy and I feel hungry. I open the van door to a sunny morning. I almost forget but there's the sign for Rivendell. (it wasn't one of my dreams) It's a great day and I feel alive and much better.

On Margaret's advice, we take a day of rest and do nothing. I agree, but can't wait till I can go and visit Rivendell. If I feel better just parking near it, what's it going to be like when we get there?

I have always felt a strong connection with Lord Of The Rings. There is a real 100% connection here in New Zealand: what with rings falling off, spider bites, then ending up in Rivendell when I am ill. I first read it when I was seventeen, forty seven years ago. It only took me three days to read it the first time. I would run home from the bus dropping me off from working the mine to get reading it straight away. I remember having to put it down on the armchair at 6.05 am and the pit bus was at 6.10 am, wondering all day at my work if Frodo was going to die. Here I was last night, wondering the same thing about myself. I was even called Frodo as a nickname from the boys. We'd go for the rock nights and I wore my denim waistcoat with FRODO embroidered on it. God knows where that went. I tell you there is a connection here. WEIRD.

MAGGIE:

After taking a combination of herbal remedies, paracetamol and antihistamines, Pete finally broke the fever. The worst was over. Pete never has to wear deodorant. He is such a calm person, he rarely breaks sweat. Even if he does, it is odourless. Right now he smelt vile, but I could see from his face he was much better. The welts were away from his body, although his skin was very dry and flaky. This was a time to slow down, recuperate and this was the ideal place to do it.

It's a lovely autumn day and I do the laundry. We end up staying here for three nights and four days.

PETE:

Next day I am feeling much, much better. We head down to Upper Hut and to the swimming pool. After having a long shower and scrub, we sit up to our necks in the bubbling jacuzzi pool. Thoroughly enjoying the relaxation and pamper, I step out and go into the sauna. I see my reflection in the glass door. I don't look too bad I think to myself. I only stay in the sauna a short time as it is making my dizzy. I feel as if something is leaving my pores. As I go back into the pool I feel as if I am healing. I see someone with psoriasis on their elbows and back and for some reason it makes me feel better for it. When I look over Margaret is quite content sitting with the odd bubble rising...

Feeling much better, we head to the real ale brewery where I'm hoping to find something to help the continuous thirst. We sit at a table inside the working brewery. It's a heavy metal themed place with Motorhead and Black Sabbath featuring. After sampling a few, we opt for the Ace of Spades. We also fill a takeaway bottle for later.

MAGGIE:

We had a spa session at the local swimming pool and both of us felt great afterwards. We then head to a local brewery and had a lot of fun tasting all the beers. We decide on Ace of Spades stout which has a taste of burnt chocolate and the consistency of treacle. For some reason, I don't know why I think of a drink I have heard of, but never tasted. Black Velvet. I look up the recipe and it says Guinness and champagne (or Prosecco if you can't afford champagne) We get a bottle of Prosecco. We're going to use the Ace of Spades since it looks like Guinness and we head back up to the campsite. There are two French couples there enjoying the evening.

We make our Black Velvets to toast our good health. It doesn't quite work. The Prosecco is supposed to sit on top of the Guinness. We wonder if it's because we don't have correct glasses. It's still drinkable. The French couples decide to party till 3am. Hey, c'est le vie!

Rivendell Bans the Ring

PETE:

The time has come to visit Rivendell and I feel like a kid at the shows (showground or fair). After all, this is where Frodo recovers from the Ringwraith's dark blade that pierces and enters his spirit. I feel so much better. I quickly get dressed and decide to wear my Robert Plant T shirt which has the face of a wolf and the words Heaven Knows on it.

The ring round my neck is off again. Only this time it's actually missing. We searched all over the van inside and outside. It's disappeared. In fact—'Invisible'. This is getting incredibly spooky, I say to Margaret. It fell off at the

Pelennor Fields, the Ford of Bruinen (where Arwen outran the Ringwraiths to bring Frodo to Rivendell to heal) and now here, at Rivendell itself. Well, says Margaret, the rings not allowed in. Simple as that.

As we enter Rivendell I'm glad there are not many people about. It makes it easier to take things in. There is a guide with two people with her. I hear her saying things like, "This is the tree where Legolas (Orlando Bloom) ponders the future of Frodo and the Ring," so I head over and stand at the exact spot and ponder my future and the ring.

The sets have all been taken down although the actual archway into Rivendell is still there. The Information Boards make you realise the enormity of the project. For instance, Peter Jackson ordered tons and tons of Autumn coloured leaves to be brought in and scattered on the ground to give the exact effect of scenes in the book.

It's a great day. Once again, I feel so privileged to be seeing all these things. I picture Frodo, Gandalf, Samwise and the others walking about. I don't want to leave, so I stay a lot longer than I expected and take it all in. One last photo of me standing next to a terrific picture of Rivendell entitled 'Early Morning in Rivendell' which has all the stairways between the misty trees and candle lights on branches. A nice photo. With that we walk back to the van, talking about what we've just seen. When we get back to the van, I open the glove-compartment to get the phone charger. There, just sitting there, in front of me, is my ring.

MAGGIE:

I don't usually dream, or if I do, I hardly ever remember them. Last night I dreamt that my dad pulled a dagger from my heart. When I told Pete about it, he says that Frodo ended up at Rivendell because he has to heal from the dark energy of a metaphorical dagger wound caused by the Ringwraiths. I genuinely didn't know that, as I am not that familiar with everything in Lord of the Rings. I don't think this is a coincidence. By this time, Pete can't find the ring anywhere as it has come off again. Then he realises that Frodo doesn't have the ring at Rivendell as Elrond has it for safety. Well that explains it then. Nothing surprises me now.

We have to walk to Rivendell as there are roadworks. It is better that way. We get talking to young Australian couple from Sydney. Turns out they had been visiting similar sites to us in the South Island. We pass a Lord of the Rings tour bus parked. Then we meet them coming down the path the guy in front is leading a dozen people. He is dressed up and has a sword. Once we get to Rivendell there are not many people. Us, The young Australians and a tour comprising of two people and the guide. They left the actual archway which is a nice touch. We end up walking through it three times, but not backwards.

Another Squeaky Bum

PETE:

Next day, we say goodbye to Rivendell and drive to Whanganui. I've got the ring back on. I got Margaret to tie it tightly. The drive to Whanganui is terrific, especially as I feel much better. According to Margaret, I look much better and smell much better as well. In fact, I feel so good that I get to drive again, which suits Margaret as she can sit back and relax. I'm looking forward to this as we are in Māori region now. There's not much traffic which means you don't have to worry about anyone driving up your arse, tooting and honking. I actually enjoy driving along, whistling and enjoying life.

According to our map we are heading towards the Whanganui River; New Zealand's longest river. It's three hundred And fifty kilometres from its starting point at Lake Taupo in the centre of the north island. According to the Lonely Planet we can expect extraordinary views at the top of the gorge looking down the valley.

The drive is far from easy, with sheer cliffs. One minute you are driving as close as you can to the cliff walls, the next minute you are looking over a drop of thousands of feet. The bends seem endless. Eventually we come to the top with a view point. Glad to see it. We stop and stretch our legs. All of a sudden hens and cockerels are running around our feet. Surely they don't belong to someone this high up? Pity they were not kiwis as we're hoping to see one. It's like trying to see a platypus: they are rare, very shy and they only come out at night.

We sit down at a picnic table to enjoy the magnificent view and put the kettle on for a coffee. A car draws up and a Māori woman gets out and lights a cigarette. She starts talking to us and I notice she has a tattoo on her chin done with black ink. She tells us each tattoo has a unique part which represents their tribe. I'm glad I met her. We are in real Māori country now. She also has a daughter and a grandson. Margaret and the woman have a real good talk and she tells her a few things of interest. It's nice, clean, fresh air way up here. The lovely looking river far below is where we are heading. It's a very steep drive, with my foot on the brake all the time: it's real squeaky bum stuff.

MAGGIE:

We head up the road, much further than we mean to. Tried not to panic at the very busy roads. Whanganui is a large town, but not as large as Palmerston North. We make it to a campsite at Sanson Motel arriving really late.

We take our time in the morning as we arrived so late last night. We end up at Whanganui East and go into a shop which happens to also be an off licence. As we were wondering what ale to buy, a song came on the radio. We can't believe it as it is a one hit wonder from the eighties or nineties and quite rare. It is called Black Velvet by Alannah Myles. We both laughed as there right in front of us were cans of Guinness. It's a sign, so we buy two cans and a bottle of Prosecco as well.

We decide we need a couple of proper glasses. Guess what? There is a charity shop two hundred metres down the road and they have two flute glasses for fifty cents each. We are all set to try again for a real Black Velvet this time.

We move on up towards the Whanganui River. The view looking over the top of the gorge onto the river is breath-taking. We are having a cuppa and writing in our journals when we notice some colourful graffiti on the cliffs across the road. One was in green and said GET A LIFE U LOVE and the other was in blue with a pink outline SPREAD THY LOVE complete with love heart. I wish all graffiti had such positive messages.

A car draws up with a gran, a daughter and a grandson. The gran lights up a cigarette and starts chatting and then her daughter joins her. They are Māori and true to form, they are very friendly. They explain a bit about the area and some good tips. This is real Māori country as Whanganui river area has actually been given back to the Māori people. Turns out she lives in a town called Jerusalem. I thought that was interesting that 'the promised land' was given back to its original people.

The river is a living spirit to the Māori people and it is treated with respect. She hears us saying Aye and smiles for that is what the Māori people say for yes too. She also tells us about the McGregor family who still farm on the land to this day. She says they don't look Scottish, but way down the line a Scottish guy called McGregor married a Māori woman. They are very proud of their Scottish ancestry and when they go to any celebrations they wear their kilts.

She also tells me of the local legend. A Māori named Hohepa Te Umuroa was exiled over to Maria Island, Tasmania in 1846. He died the following year of tuberculosis. In later years, his family went over to collect his bones and brought them back to this area to be buried at Hiruharama (known to settlers as Jerusalem). A bird called Nankeen Night Heron, which is native to Tasmania, appeared in the area shortly afterwards. It is believed that they carried his spirit back home over the sea and still live in this area as a kaitiaki or caretakers. It's the only place in New Zealand where they are found.

I mention that the Māori language reminds me of Native American and they start to sing a traditional song for us called When the mountains meet the sea, only they sing it in their language. It sounds so beautiful. What a privilege to speak to these women.

The River Queen

PETE:

Well, we made it. Man, that was some drive. I feel I can drive anywhere now between this and the Lion Road. We're at the bottom of the gorge. We drive along the river for a spell and pass a few places of interest. Matahiwi had a little house and wooden cabins with canoes bobbing about tied up with ropes. As I looked up at the only clearing in the opposite hillside I saw two horses. They must be wild I reckon. Do you get wild horses in New Zealand?

Then we come to a small place called Ranawa. The few buildings were definitely Māori in structure with carved tiki sticking out their long tongues and fierce looking eyes. That's to scare away the evil spirits.

There's not a soul about as we wander round the three buildings. They are painted white with maroon coloured centre posts and railings. I think one of them might be a hall of some kind or a church even. Maybe the tiki have frightened everyone away. It's got a real, nice friendly feel this place. Much friendlier than the South Island. I think it's the Māori vibe.

We move on and come across a sign which says Campsite Welcome: Open and in we go. It has a big lounge with settees, cooker etc. A real feel at home vibe to it. We decide to stay.

There's a big coloured poster on the wall, but I can't make out the words. When I inspect it, the writing is all mirror writing. It says River Queen and it's got a picture of a big canoe with five people in it: a white woman; a white man and four Māori men. I recognise the white man's face but the actors names on the poster are in mirror writing as well. His name is…what do you call him… He's got a famous father that I like as well…in fact he's got a band as well albeit a little known one… I remember his father being in a load of films…a spooky one about bodies being taken over… I look at the mirror writing again. It starts with a K Kaf Kif Kie. Got it. Kiefer Sutherland and his father is Donald. The poster looks great. I decide I want to see it.

I am looking at it again in the morning when a pleasant Māori woman comes in. The poster you are looking at, the photo was taken out the back here. They filmed the River Queen here she says and we got this poster as a memento. It's about a white girl and a Māori breaking customs and causing upset between Māori tribes and British settlers. The caption at the bottom says:

A kidnapped son

A nation at war

An Epic Journey

All the scenes in the river were filmed right here. She looks well chuffed. That was the river we were looking down on and saw the canoes tied up… As I like Kiefer anyway, I make a mental note to try and get a hold of the film. From what I have seen and heard, I am looking forward to it.

MAGGIE:

We just amble along the road next to the river. The cliffs seem to be crumbling as there are lots of rocks lying on the road. Think this could be a daily occurrence. There are lots of beehives. Think this could be prime manuka country. We stop at a place which I would describe as a fairy glen and take some photos.

We carry on up the road and stop at a Marae which is beautiful and then it's on to Jerusalem. We stop to look at the church. Some beautiful pictures explaining the rosary standing for the mysteries of life. Here was me thinking the rosary seemed to stand for penance and guilt. Then, there are lots of rosary type rituals used all over the world for different belief systems.

I sat on a seat and notice it says Trust in Jesus. I ask Pete to take a photo and my phone goes out immediately afterwards. I realise that this is the truth. I just need to trust in Jesus in my heart, not my mind. That is enough.

Keep on going to Pipiriki as there is a campsite next to the river which cater for white water rafting and canoeing. We park up as it is dark. When we go into the lounge/kitchen area, there is a French couple there and the woman is quietly knitting. I point out a cuddly toy wolf pup to Pete and he immediately calls it Mountain Mist saying it's a relation of Mountain Cloud from his dream.

Time to try our black velvets. This time it works. Pete does a great job. The Prosecco is sitting on top of the Guinness. It's the same sort of idea as making an Irish Coffee except it is cold. It looks a treat and tastes pretty good too.

A very pleasant Māori family own the business. We find out the film River Queen with Kieffer Sutherland was made here. Pete was well impressed.

There are lots of giant silver ferns here and they are the size of palm trees. They make the woods look like rainforest. The silver fern is a national emblem of New Zealand along with the kiwi and we are told that kiwis are known to be around the Pipiriki area. We really enjoyed our chat and our stay.

Get Your Kicks on Route 66

PETE:
We leave and end up driving through the Tongariro National Park. I am thinking it doesn't get much better than this as I'm feeling a million New Zealand dollars. We've not been driving that long when Margaret stops the van. Picture please she says jumping out the van.

I wonder what she is on about. Then I see when she walks a bit, there is a sign CAUTION: CROSSING AT NIGHT and above the writing there is a picture of a kiwi. I know what she is thinking, the same thought when we saw a koala sign in Australia. We did end up seeing a koala in the wild.

Further along the road there is a terrific view of Mount Ruapehu, the highest peak in the North Island and another location for Mount Doom. We see a signpost with the brilliant name of Whakapapa I instantly take a photo to show my grandsons.

As I am taking the photo I remember the atlas saying something about Mordor. We drove up to the top. We get out the van, walk across the park. We look out to a landscape of molten lava rocks and craters looking like my idea of the moon. I jump onto a rock and instantly picture Frodo and Samwise at Mordor. I only walk a few steps and the flipping ring has fallen off AGAIN. I get Margaret to take a photo of me picking it up. It has certainly taken on a life of its own since we have been visiting these locations.

There are over one hundred locations. Some of them are only accessible by helicopter. In the distance is another view of a mountain top covered in snow. It's easy to transport yourself to Middle-earth here. In winter at Whakapapa, the rocks we are standing on are covered in snow and it becomes a playground for skiers. We leave Mordor and me wondering what on earth is going on with the ring?

We drive on and come across a town called Turangi. There is a cafe with a huge pink Cadillac made out of corrugated iron with the words cafe on it. It's hard to miss actually. We want to go in for a snack. We are disappointed as it is closed today. Outside there are loads of these 1960s American metal posters Harley Davidsons, Volkswagens, Women and of course the most famous of all We Got our Kicks on Route 66. To show off I take out my bag, given to me by my daughter Elizabeth. A real Route 66 one with all the different jargon written on it. The obligatory photo was taken.

We pass a place with the totally unpronounceable name of Te-Rangitautahanga. It's a nice wee place and we stop awhile. Then off to Two Mile Bay where we stop to see another beautiful sunset.

MAGGIE:

We are heading up a long straight road towards Lake Tapua. The mountains, river and lake were formed by a volcanic eruption. The largest in the last five thousand years. It's ashes were seen in the sky as far away as China and Europe. Today it looks like another taste of paradise.

We see the still live volcano in the distance. I see something which makes me stop and pull over. My first yellow sign with a picture of a kiwi. That makes me feel so happy because it means that there are enough of them here to actually need to put a sign up. It was like seeing the koala sign in Australia. Both iconic creatures.

Pete is happy to take pictures of Mount Doom volcano in the distance.

We also see a sign for Whakapapa and decide to give it a visit. From this small ski resort, we are able to drive up and see the Volcano up close.

We drove on towards Matamata. We hadn't planned to do this. It just happened spur of the moment. We didn't expect to drive so far.

PETE:

We don't stay there as we have to be at Matamata in a couple of days. We have to get Noah's sliding door fixed at a garage there. It's quite late, it's dark and we're a good bit away yet. Our phones are out because we have no cable and so there is no sat nav.

It seems like we are going round in circles and there are lots of possums out on the road tonight. We go into a town and park at a campsite next to a sports club, but we don't get a good vibe from it. As if to prove it a sign saying DANGER is stuck on a fence, so we follow our intuition and move on and find a better place. Eventually.

It is dark and we have to rely on our senses. This place feels OK. Straight to bed and up early. When we rise, we discover our senses were correct. It's a nice wee place. As I am cooking breakfast a tourist minibus arrives. They seem to think we're cute and all start taking photos. Fame at last.

MAGGIE:

I am beginning to panic as petrol is running low, we have no signal or charge on our phones. We are in the middle of nowhere looking for a campsite in the dark. I try to keep calm which is the best thing anyone can do for themselves. Keep calm and slow down. We went into a town and there is a campsite, but I have an uneasy feeling. I don't know why. I say to Pete and he says he agrees as he just saw a sign saying DANGER. I am getting better at listening to my intuition and trusting my instincts. It ends up we camp at a picnic area with toilet. This feels right.

Wake up to another beautiful autumn morning. Feel blessed. A mini bus full of tourists pulls up and take photos of our campervan and Pete making the coffee on our wee stove. We look at our atlas to get our bearings. We are near Putaruru which, thank goodness, has a petrol station and is quite close to Matamata.

When I woke up, the thought crossed my mind that I haven't been to a New Zealand Church service yet. As we drive into Putaruru, Pete notices a sign for a gospel church. I swiftly take a right in search of it. Pete doesn't feel the need to go in the church. He says his faith is strong enough he doesn't need a church to make it stronger. Sometimes I like the communal thing. If it is uplifting, if you can feel positive energy. When I find the church, the service has already started. Just as I am about to turn and leave, a young family arrive late and invite me to go in with them.

It's an interesting sermon based on Luke Acts: Jesus is in everything and is in ourselves. He is our hero not us. Do our 'thing' that we came here to do and thank Jesus as you do it. I am willing to let Jesus in my heart. Only I can do this.

The Pastor comes over to talk to me and asks where I heard about the Church. I said my husband saw the notice, but he doesn't need to come in as he feels his faith is strong enough. I think I gave the Pastor food for thought, just like he had done for me. Jesus is the hero not us. That is exactly what Pete believes, he doesn't need to hear the sermon as he knows Jesus is his hero not the Church. I gave thanks and made my way to meet my sweet Pete.

PETE:

Feeling good about our claim to fame we finally find a town with a petrol station. Putaruru. As we fill up I notice a sign for gospel church. This suits Margaret just fine. She had been worried about running out of petrol and getting stranded. Now that the van has got juice in it she can relax again and join in some good old gospel. She hasn't done it for a while and so I am happy for her. However, I am not going. I don't feel the need as my faith has been strong since I was a teenager. I also still feel a bit zapped with my own spider bite juice and this ring thing is doing my head in. Why does it keep falling off? In Mordor of all places. As if it is inching it's way home? LOL?

Cambridge — Ma Baker

PETE:

Margaret's feeling a lot better and enjoyed the sermon. She said, the gospel music wasn't as good as the Australian ones. After my walk in the fresh air, my head feels better as well.

So feeling happy, we drive on to a place called Cambridge. I like Cambridge, the one in England. I know it well, as my mum lived there for years, remarried and became Mrs Baker. I've been there a few times back and forth, over the years often hitch hiking with a friend and arriving unannounced. My twin brother stayed there and worked there for a few years and he would bring friends with him too. My mum became known as Ma Baker. We just had to go and visit the New Zealand Cambridge.

I am surprised as we drive in to see a sign saying Victoria Station which is the same name as the station in the other Cambridge. It turns out there are many similarities between the two Cambridges. There are push bikes and a cycling map and loads of other wee knickknacks which remind me of my time there.

My mum passed away last year and this is bringing it all back. I can just visualise her laughing, with her ginger hair. I turn the corner and a huge billboard in front of me says MA BAKER. It's a baker shop, but I don't half get a 'stopped in my tracks' moment as I was just thinking of her. Margaret does her wise woman of the north thing and suggests that we sit down somewhere get a coffee and think of her. I'm glad Margaret got to meet her and we drink to that as well.

After our coffee, we stroll past Ma Bakers again and there is a recycling clothing boutique with a great name FROCKWORK ORANGE. We have noticed that the New Zealanders like their play on words. I think it's great.

We decide to go to the Cambridge Heritage Centre. There is a lot of pictures and information on the English Cambridge with all its museums and colleges. There in the corner is one thing that does catch my eye and it has got nothing to do with Cambridge. It's a wooden beer barrel, with glasses of tasters on top, clay pipes you can smoke and the words on the middle of the barrel: SOUTHFARTHING HOBBIT ALE. We are in Hobbit Country!

Hobbiton is the next step on the road tomorrow. I am all excited as it is my birthday tomorrow and as a present Mrs Hobbit has already booked us in.

MAGGIE:

We decide to go to Cambridge which is only thirty kilometres away. Great vibe. We see Victoria Street and Victoria Station. Pete is thinking about his mum. There is a giant Ma Baker sign across from the Town Hall, which I think is appropriate as she worked at the Town Hall in Cambridge. We toast our own Ma Baker with a nice coffee.

We are heading for Matamata next. I am excited about tomorrow, but I know Pete is way more excited than me. We found a great campsite for six dollars a night. Ideal. Again, the town has a good vibe. The people are friendly.

The alarm gets us up early. Before we visit Hobbiton, we are taking Noah the campervan to get the side door fixed. The door is sorted within twenty minutes. We noticed an auto electrics earlier and after speaking to our van hire we get the go ahead to get the radio/cd player fixed. We end up with a new one which plays Spotify, charges phones and does Sat Nav. All at the same time! Happy Birthday Pete. This means he can play Led Zeppelin for the rest of the day till we go to Hobbiton.

Hobbiton — There and Back Again

PETE:

The day has arrived. It's my birthday and we're going to Hobbiton. I'm not the same age as Bilbo Baggins. He was 111 on his birthday and I'm only 64. So I get to sing the Beatles song When I'm 64 for real and it's only apt I put my Sergeant Pepper T Shirt on.

First, though, we have to get the van fixed. We get the door fixed as well as the electrics and now we have music, a sat nav and phone chargers. We explore the town of Matamata first and there is a Hobbiton Information Centre all done up like a Hobbit house. Very cool looking. As you walk in the first thing you see is quite a spooky looking statue of Gollum. This makes me feel round my neck to see if you know what has fallen off. It hasn't. I give it a tight tug and it's secure enough.

I'm delighted to notice that lunch is included. In the book, the Hobbits love their food. In fact, they can have as much as three breakfasts just deciding what to have for lunch, never mind dinner. I'm well impressed.

Well, says Margaret, you better jump in if we are to make it in time for your birth time. Don't you mean birthday I laugh. No she says look at your ticket. When I look, it says HOBBITON MOVIE SET TOUR AND LUNCH COMBO 6 MAY 1:20 pm. That is the time I was born 1:20 another nice coincidence. Margaret's good at that kind of thing.

It reminds me of our Wedding Day on St Valentine's Day at Burleigh Heads. These little touches. The previous year, as we walked at the beach at Greenan, Ayr on St Valentine's Day, I found a stone. It said 'Be Mine' with four love hearts pegged on a washing line. The same picture on our calendar for that month and with that thought it's off we go to Hobbiton. TOOT TOOT HONK HONK.

Put Hairs on Anyone's Feet

PETE:

We're here and there is a definite feel of excitement. Our Guide checks our names, the time of the tour, counts some heads and off we go like school children on a bus trip. The actual set of Hobbiton is a twenty minute drive from the car park and the Guide takes the opportunity to fill us with all sorts of unknown facts.

How part of the sheep farm was transformed into the Shire. The road which allows access into Hobbiton was built specially by the New Zealand Army. He informs us that Peter Jackson, after finding the location, getting permission from the farmer, actually built Hobbiton a year before filming. This enabled gardens to be established, vegetables to be grown for a more natural and lived in effect.

As we arrive at the Movie Set, we get off the bus and to my delight there is a big tent. Right, says our Guide, food and plenty of it. At the banquette, some of the folk are really up for it Gandalf hats, Wizard staffs. There's even a guy with a beard, curly hair and bare hobbity looking feet, then I realise I am looking at my reflection. There is everything you can possible think of. I not only look like a Hobbit, but I can eat like one too. I have a piece of everything and my plate's brimming. I leave room for pudding as after all it is my birthday and I wash it down with a flagon of gooseberry wine as well. Then the tour begins.

It's just what you expect. Full of paths, with wooden picket fences and gates. Flowers in all colours leading up to round Hobbit holes with round doors all built into the rolling hills. There are full time gardeners employed to keep the place looking tidy. What a job I think. What do you do? I'm a gardener at Hobbiton. Love it.

As you wander round there are lots of photo boards showing what it looked like during filming. The Guide tells us, "Do you know the bit in the film where Gandaph hits his head on the wood when turning round to face Frodo. That wasn't in the script, but when it happened they kept it in." As we walk further there are rows of little Hobbit size clothes hanging out on the washing line. There's smoke coming out of the chimneys. You honestly think Bilbo Baggins is going to come out of the door. They do a great job here. The Guide also tells us had we been here a couple of weeks ago Harrison Ford requested a private tour, but since he was only allowed to go twenty minutes before everyone else, the next tour came across him.

We head up Bagshot Row to Bag End to see the most famous Hobbit Hole in Middle-earth. First the Guide points to an area just before we enter Bagshot Row and says I'm going to take you up to where Bilbo held his famous Birthday Party. Then he stands under a big tree and disappears saying can any of you see what is wrong with the trees on the road up? I can't see anything wrong and being a Herbalist I know my trees. Margaret, who was brought up on a farm, can't see anything either. Nobody does.

I do notice there are bee hives here. When we get to the top, the Guide asks if anyone has worked it out. Well he informs us this tree got split in a storm and as it was such an important part of filming Peter Jackson got the prosthetic team to build one. It's not real. I am amazed because it looks as real as the next real one. The Guide goes on, "And when it came back and was put up Peter Jackson said the leaves are the wrong colour for the time of year it was supposed to be. The workforce had to take all the leaves off by hand and replace thousands of them to the colour he wanted. That is how much attention to detail went into the making of this film."

With that said, The Guide turned round and said to everyone this is where Bilbo held his 111th Birthday Party so he says let's sing Happy Birthday to Pete. It's his birthday. Imagine. I'm standing in Hobbiton, at Bilbo's tree, getting Happy Birthday sung to me. That is a birthday treat I will never forget.

We then walk past the tree onto the path with lanterns hanging from the branches and walk up Bagshot Row to Bagend House. There is the famous Hobbit Hole with the famous sign on it that I completely forgot about NO ADMITTANCE EXCEPT ON PARTY BUSINESS. The Guide says to us to stand at the Hobbit Hole and he will take our photo. We are the perfect size and the photo is so good we both have it as our screen saver on our phones. I honestly wish I could show you them. We were tailor made for a photo at a Hobbit door.

It's time to head back to the Green Dragon. The Guide takes us over a hill and past the old Mill, It's the path that Gandalph and Frodo are in the carriage as they plod to Hobbiton. Finally we go into the Green Dragon and a log fire is burning. There's a glass of Southfarthing Ale with the words 'put hair on any hobbit's feet'. Bilbo's red waistcoat is hanging up and I put it on and it fits a treat and get another photo taken. The guide gets the drinks in then comes over with our Southfarthing ale. Well that's you done it. YOU HAVE BEEN THERE AND BACK AGAIN.

What a great day. I say a special thank you and farewell to our Guide and tell him that I won't ever forget this birthday. Thanks Margaret.

MAGGIE:

It really was a special day out to Hobbiton. Witnessing Pete's pure delight in seeing his favourite books and films coming to life before his eyes, gave me such a wonderful thrill. It is evident that for Peter Jackson it was a labour of love and it is evident that for Pete, it really is the realisation of a dream to come and witness it.

There were approximately 650,000 visitors here last year, twenty years after the film was made. It's A Wonderful Life is my favourite film of all time and I know how I would feel being on the set of that iconic film. Pete just loved being part of it. He looks like he was born to come here with his wee hobbity bare feet and shorts on, his curly hair—all five feet three of him. He really looked the part. In fact, we both looked the part standing at Bilbo's door. He still can't get over being at the same spot as Bilbo and folk singing Happy Birthday to him on his birthday at that very spot.

I want to say our heartfelt thanks to our considerate, patient and knowledgeable young Guide. He also told us that usually forty to fifty folk go on one tour. To have only eleven that special day meant there was more interaction which made it all the more memorable.

When we got back to the van, we had a karaoke on Noah's new music system. We played and sang When I'm 64 by the Beatles and then I chose a song I know is dear to his heart. It meant such a lot to both him and his mum: Simple Man by Lynyrd Skynyrd,

Simple Man

PETE:

That night we sit in our van. We can listen to music again and since it is my birthday we have our own wee karaoke. We sing When I'm sixty four and Simple Man which always makes me think of my mum. She used to give me that kind of advice. To this day it still rings true.

We don't wake up to a glorious sunshine morning. In fact, it's twelve o'clock in the afternoon and to my dismay my welts have been there and back again as well. At least, I was well enough to have a great day at Hobbiton. I make a decision to not even think about welts. I am having a great time and if you travel to different countries then you should be prepared for what some people have to put up with on a daily basis.

As we are still in Matamata we wander round and everyone starts to look very familiar. They all look like extras out of The Hobbit. I'm not kidding. I bet half of this town were in it. We feel very much at home LOL. I notice one of my favourite shops a charity shop. It doesn't matter where I am there is always one.

In we go and as I walk around I see a Marino Wool top. They are usually eighty dollars and this one is eight. I am an expert on charity bargains. Margaret shakes her head in disbelief. How do you do it? she says.

Got Gortex walking boots in Brisbane for three dollars; Abercrombie and Fitch shirt for one dollar; a good pair of denim jeans for fifty cents and that's me now rigged out from head to toe all for twelve dollars and fifty cents. That's the way to do it.

In the same shop, I also notice a thick leather thong. As the ring didn't fall in the fire at the Green Dragon, I decide that now is the time to get a more sturdy leather thong. So I buy it for a dollar. It's got something on it, but I like the look of the thong. It looks perfect. I take off the circle shape and put it in my pocket. I slip the leather thong through the ring and tie it. This time it feels very secure and dare I say it 'lighter'.

As I head to the van Margaret asks what was on the thong when I bought it? Did you bin it? I don't know what it says I've got it in my pocket. I bring it out and it's a silver looking flat circle with a hole in the middle. It looks like a decorated washer. It's definitely a piece of cheap jewellery. There's writing on it. We turn it clockwise to read it and it says BE TRUE TO YOURSELF. I just shake my head. Nothing surprises us anymore. In fact, I still have it. It's in my mojo box with a lot of my other weird finds.

MAGGIE:

We had a restless night and then slept in till noon. Pete woke up with more hives on his side. He is determined it won't stop him enjoying himself. He certainly doesn't feel as ill as last week. The worst is over. By the time he has been to a couple of his favourite charity shops, he has plenty of bargains. He looks like a rock star with all his designer gear.

We are heading to Rotorua. It smells like sulphur which is the ideal place for me at the moment because I do too. All night and all day, the last twenty four hours. To make matters worse, walking back to the van, a lady handed me a leaflet recommending an Indian Restaurant. I was so hungry I couldn't resist. It will either kill me or cure me.

I woke up my heart was racing and I didn't feel like moving. I must have some sort of virus and that's what has been wrong with me. We had to be out for 10am. Pete has taken over the driving. I felt worse as the day went on.

We ended up at Gisborne. I don't remember much about the journey. A lady recommended a hot ginger wine toddy. We camp at the beach in a DOC campsite outside the town. I take her advice. Hot ginger wine, echinacea, golden rod and two paracetamol. I go to sleep listening to the waves which are very soothing. It does the trick as I sweated it out and I feel much better when I wake up.

Snakes and Ladders

PETE:

When I wake, I am feeling so much better, even if I don't look it. My positive thinking is working. My insect lotion still needs something else to make it more longer lasting. From now on, I am going to keep focusing on insect repellent.

I turn round and can't believe my eyes Margaret is not looking too great at all. Surely she has not been bitten. I'm nervous at the thought of it. Turns out it is some sort of stomach virus bug. The tables have turned. It's like snakes and ladders. Margaret is unwell and I need to drive. I am so glad we weren't unwell at the same time. That would have been a nightmare.

We are going to Gisborne to experience the first light of day on Earth. Looking forward to that alright. Think I will sing Morning has Broken when I see it happen. Now that would be cool. First though, we need to get there. To do that I need to drive through the Raukumara Range. I've had two squeaky bum drives so far and I'm thinking I won't be disappointed with a third.

Margaret has taken some echinacea and golden rod. We think it is a virus and that's just the chap for the job. She also takes some paracetamol to help sweat it out. I start to drive with Margaret lying in bed in the back, like a patient in an ambulance. It feels strange not having her to talk to as I pass all the wonderful sights. I think that this is how Margaret must have felt with me last week.

I start getting the hang of it. I begin to feel OK that it just me in the front. There is a sign Raukumara Range and it might as well be saying Welcome to a Squeaky Bum Drive.

It is pure concentration all the way uphill with narrow roads, sharp bends and cliffs. You could go round a bend and go straight into a steep hill at the peak of the bend. If I were on skis, I'd love it. It's a quick turn and a quick descent. Then, as if to give me a break from using my brake, I get to put my foot on the accelerator and it's squeaky bum time. You think you're only inches off the sheer drop as you brace yourself for every turn. Will there be passing places? Is there enough room in the passing place? There's also the odd bundle of debris so avalanches spring to mind as well. Every now and then there is a row of forest trees as your outside rail, which is kind of OK because if you do skid off at least you are hitting a tree and not tumbling down a ravine.

Every five minutes or so the tree reveals a gap to remind you how high up you are. Then, as the road finally widens, you get the feeling you are on flat land again. Relax for five minutes then it's squeaky bum time again and 100% pure concentration. A few hours later we are on State Highway 30 as we leave the Bay of Plenty and Raukumara Range and Gorges behind and we arrive at Gisborne.

Margaret is still not feeling too good, but we go for a walk to stretch our legs. We get talking to someone who recommends a hot ginger wine toddy. We take the advice, after finding an area next to the beach just outside Gisborne and we camp for the night.

When we wake, Margaret feels a lot better. The virus has been broken because of the complimentary medicine combination of hot ginger wine toddy, echinacea, golden rod and paracetamol. In fact, she says she feels well enough to drive again if I want. It's a nice coastal route with nice fresh air and flat roads.

MAGGIE:

We found out yesterday that not only is Gisborne a town, but it is also a district. The East Cape Lighthouse, where the first sunrise of the new day arrives, is a fair drive up the road. We have also been told it's Māori country which pleases us. Suddenly, we come to a police check and I am asked to produce my driving licence. The policeman asks why we are going into this area? I explain we want to go to East Cape and tour round the whole peninsula. He seems satisfied and we are on our way.

It's a beautiful drive along the coastline. We know that the longest wharf in the southern hemisphere is around here. Once we find it, it's well worth a visit. As we walk towards it we pass through a Māori influenced tiki, the wharf feels quite dark and shaded to begin with, but in the distance there are blue skies and sunshine. Sure enough our effort is rewarded with crystal clear blue water and blue sky.

As we drive on, a couple of towns we come to feel down on their luck. That is nothing new to us though. Back home, Ayrshire is full of these types of villages: old mining villages which have seen better days and lost their self-esteem. In the Māori villages, it is pubs and hotels which are closed down, even although the beaches are some of the most stunning beaches we have seen in New Zealand.

We stop and go into a shop for an ice cream to soothe my sore throat. I asked if it was always this quiet. I was told by the friendly lady serving that there was a funeral on today and most of the village were at it. I saw the flavours of the ice cream and again it reminded me of home in that they were extremely sweet. There was Rainbow Hokey Cokey and Goody Goody Gumdrop. I plumped for a two dollar scoop of rum and raisin. It was the biggest scoop she could manage with enough raisins to make a clootie dumpling back home. So generous. So delicious. Thank you.

Kept on going until we stopped at Tiki Tiki to see the stunning Māori influenced St Mary Church with its quaint crocheted cushions sitting alongside exquisite Māori carvings. It had a great vibe. I said a prayer in front of the stained glass window depicting two soldiers in uniform kneeling before Jesus on the cross. I thanked both the soldiers and Jesus for giving the ultimate sacrifice, so we can all be free.

We moved on to Te Arolea and found a backpackers campsite which had a Zen like atmosphere. It's time for an early night and another hot ginger wine toddy.

Carry on Driving

PETE:

The reason we are driving along the coast all the way to East Cape is that we find out the first light of day on Earth can be seen up there. It's in the district of Gisborne and not the town of Gisborne. It's a long drive and a nice day for it. We check our Lonely Planet book and find out that we are driving into real Māori country where hardly any non-Māori people live. We agree that this is more like it. This is what we want and not always touristy things. We head off into the unknown.

It's a nice type of Sunday afternoon drive and we come across the longest wharf in the southern Hemisphere. We have seen loads of interesting things on our journey and this is no exception. It's a long way out to the end of it. We go out, stand and admire the vast blue sea. When we go back to the van, we make a tea and admire the wharf from a different angle, before heading on.

I think the sea air has made us both dry. When we come into what looks like a one horse town, we stop to get an ice cream or fizzy drink. It's not hard to see that it is quite rundown. I notice that one of the buildings has the word HOPE painted on the side of the wall in huge yellow letters.

The place looks empty. We find a shop and a Māori woman comes to the counter. She has a tribal tattoo on her chin not unlike the friendly lady from Jerusalem, Whanganui. I resist the urge to ask her what tribe she's with. I don't know why because I would have loved to have asked her. She is very pleasant as we find most Maoris are.

Just like the Aborigines in Australia I feel like they have had a real bad time of it. Both cultures are fascinating, but the introduction of alcohol, drugs and sugar to their lifestyles has certainly taken its toll over the years. She tells us that yes there is a sacred piece of ground up the road and up the field a bit which is listed in the Lonely Planet. To be honest she says, there's only a post left, so really, it's not worth going.

In the shop, we notice ice cream with half a dozen different flavours. Just the job for my dry mouth. We both get rum and raisin. I can hardly hold it with one hand. It's massive with a capital M. There's got to be a pound of raisins, plus it's rummy. Margaret comments that when it comes to giving generous helpings it is usually the ones with the least who give the most. As we drive away, I can't help but look back at the wall with the word painted in big, yellow letters: HOPE.

MAGGIE:

Although the weather forecast it not good, we still manage to sit out and have our breakfast in the lovely zen garden. Pete got talking to the owner he said he had Scottish ancestors and loves Scotland. Pete wrote a message on the wall along with hundreds of other visitors.

We find that Te Araroa is a great place to base ourselves and explore the East Cape area. There is a pohotukawa tree next to the school. It is reckoned to be six hundred years old and the oldest pohotukawa tree in New Zealand. It stands sixty five feet tall but it's girth is the most impressive one hundred and twenty feet. To this day it is looked on by the Māori people as sacred and to have great wisdom held within its branches. No one wants to be the one to cut any branches off and disturb the Tapu.

174

We see a sign for Manuka Honey and think it might be a stall. We often see these stalls all over Australia and New Zealand with local farmers or enthusiasts selling their seasonal produce such as bags of oranges, avocados, or jars of honey, usually with an honesty box. I like that the trust is there and no one wants to break this trust. Honesty boxes are in lots of campsites as well.

What a lovely surprise. It is actually a large scale Manuka oil and honey producer with a coffee shop. Torrential rain comes on just as we arrive. We get talking to the owner and also her husband when he pops in for his coffee break. We are so interested in what they do here. They have taken a pure, natural substance and harnessed it's phenomenal healing properties and now sell it all over the world. Similar to two people who have inspired Pete over the years: Alfred Vogel and Jan de Vries. As we both are very interested in natural healing we could have talked for hours and we did. If we had come in the high season, it would have been too busy to talk as busloads of people come here to visit.

Pete explained what had happened with the spider bite and she says she has heard of people having these bites and it is known to cause people's skin to actually rot. (It certainly smelt like that to me) We realise how lucky Pete has been although it didn't feel like it for him at the time.

We asked for two bacon rolls and two flat whites. Just like the lady with our rum and raisin ice cream cone earlier, she went the extra mile. Best bacon rolls we have ever had on lovely ciabatta bread rolls.

We found out that the owners both had Scottish ancestors: a great grandfather married at Māori woman way back in the whaling days. Similar story with her husband and the girl working in the cafe with Sue had Scandinavian ancestors with similar story.

We had a great conversation about the differences between the North and South Islands and about racism. The people in the North are so much friendlier. Sounds familiar I think.

We bought some products for Pete's skin: soap, pure manuka oil, moisturising cream and an immune system boosting tea. When I opened the bag, later the lovely owner had slipped in a box of antioxidant tea and a tube of manuka cream as a present. Lovely people.

Hope – A Lighthouse Keeper's Beam

PETE:

Once again, it's been a longer drive than we thought and once again it's late. We wake up with our van neatly parked at the edge of a zen garden. As I walk round I get a good feeling. It's like 'Peaceville' man with objects, flowers, water lilies floating in pools and little statues of animals and Buddha. As I walk into the communal kitchen the owner walks in. He is Māori and as we start talking he says you're Scottish, I can tell straight away as I have Scottish connections.

He tells me his name and says he loves the Scottish people. There are a lot of Maoris with Scottish connections he goes on to say. He hands me a marker and asks me to write my name on the wall. I look at the wall. There are lots of names with the country they are from, so I write down Pete and Maggie fae Scotland. Aye he laughs and with a try at a Scottish accent he tells me a bit about himself. I laugh GID YIN and then head to the van. With a toot toot we are away.

We stop at the beach and start to walk. There's a Church up the hill a bit so we head up that direction. It's got Māori carvings on an archway, which you have to walk under on the way in. The Church is open and it's got flowers all around. It's obviously still in use. It's very clean and it's very old. I can tell by the wood. The carvings are great. There are lots of figures of men and women. The men have got their tongues out and the women have their, well, breasts out.

There are little cushions on the pews with bibles sitting next to them. I pick up a bible and with my thumb slowly flick the pages like a pack of cards. I had to look twice. It is written in native Māori language with the English Christian equivalent on opposite pages to each other. I would have loved to hear a sermon in their own language. This is real Māori land alright.

There's some very old gravestones in the surrounding garden with small wooden boats lying next to them. There's also a few newer white gravestones also with objects beside them. I get a feeling they are sacred, so I don't go in. I walk back down the carved archway and out onto the road again.

We are visiting an old tree. I see the huge multi branched tree. It's like a group of octopuses sitting together. I've never seen so many branches on the one tree. There's a sign next to it, but at the moment it's too far away. I'll need to walk round the tree to see it. It's sprawling branches are vast, fat and long. I can't resist climbing on it. I climb over and in between branches for a while and jump down at the other side. Not bad for a sixty four year old I say to myself and land next to the sign DO NOT CLIMB ON THE TREE. I sincerely apologised to the tree. I am glad no-one saw me. I read the smaller sign above it. It's called by its Māori name Te Waha O Rerekohu and the school where it sits is called the same name.

I also notice there are very young children running around at the far end of the playground. There is a man doing a lot of running around. As I got closer the man is actually rolling car tyres at the children. The children are running towards the tyres and they are knocking them down with their shoulders. I realise that they are training for rugby. I also realise I was lucky to get off the tree when I did. A tour bus drew up and a bus load got off taking photos. The driver was an official looking guy in a uniform. Time to move on.

We see a sign saying Manuka Honey. We drive down and just as well as it is starting to rain we see a sign saying Honey and Cafe and decide to have a late breakfast. Two bacon rolls please.

The cafe/shop sells everything made from the manuka tree: oil, honey, soap, cream, dried tea leaves to name some. Tea tree looks more like a bush. It's white flowers are irresistible to bees. We all know it is very, very expensive back home. Here in New Zealand you come across local farmers and hive keepers and can get a bargain. We find out that this place is an actual factory and it's been producing these products for the last thirty or so years. It exports all over the world, especially to Germany and Japan. These products are 100% genuine manuka. I got speaking to the owner and told him I was a herbalist and we had a real interesting chat. He was interested in the story of Chief Black Elk and the snake and how it lead to the discovery of Echinacea. As it was a working day he couldn't wait too long.

Our bacon rolls were the best we've had. I told the owner who runs the shop and cafe, about the spider bite. She said that I was very lucky. She knew of cases where the skin had actually rotted away. That's when Margaret told me that I actually smelt like death. Eh! Well I'm sure glad she didn't tell me that at the time. We bought a few things and we noticed later that she had given us a present of a tube of manuka cream and some antioxidant tea. There were no welts on my skin, but it was very dry and scaly. Boy was I lucky.

MAGGIE:
It's afternoon by the time we leave and head up towards East Cape Lighthouse. The rain is easing off. It is a long and winding road. We deliberately

take our time. When we get to the Lighthouse, it looks a very steep climb up to the top. I am still not feeling 100% well, so I decide to sit in the van while Pete climbs up himself. When he comes back a good hour later, he says he is really glad he did it.

This is the most easterly point of the world and therefore the first place the sun rises. I am really, really wanting to get up to see this and will need to rise at 6.15 am. Early night for us and I take my medicine including drinking my new tea. Sweated it out again. In the morning, I feel able to get up no problem.

So this is what the word AWESOME was invented for: to describe the feeling I had seeing this. A feeling of gratitude and being blessed to be able to do this trip came over me. HEY it's a brand new day and a fresh beginning.

Pete drove back along the road we arrived on yesterday. It looks so different in the sunshine. We have been so lucky once again. I see a kingfisher sitting on the telegraph wire. Then another one a bit further on and then another one. They remind me of their cousin the Kookaburra and I smile.

PETE:

Our new friend tells us that we can drive up to the Lighthouse which is not far from the spot where we want to go and experience the first beam of sunlight to reach our Planet. She warns us the drive the Lighthouse is not for the faint hearted. No surprise there then! The drive is indeed not for the faint hearted.

At first, it starts off running parallel with the beach, then the beach disappears and it's rocks, then the rocks disappear and it's road like no other. It reminds me of those roads you see way up in the Andes where everybody has to balance or you are over the edge. As we slowly nudge our way forward there's a sign just before us that says EXTREME CAUTION with pictures of boulders hitting a car that's on two wheels, with the other two wheels up in the air. We drive on, very slow. If another car comes, we are in Shit Creek so to speak. We keep going until as we turn a corner we see horses in a field and there's a farm. We've made it.

It's quite spacious but whoever owns this must be half goat to be able to live among these rocks and this road. We can't see the lighthouse, but an information sign tells us THIS WAY and further up the sign says EAST CAPE LIGHTHOUSE and then another sign TO GO TO THE LIGHTHOUSE YOU HAVE TO GO THROUGH THE FOREST IN AN UPWARD SPIRAL ON THE WOODEN STEPS THAT WERE BUILT. THEN THE LIGHTHOUSE IS AT THE TOP. THERE ARE 602 STEPS. Margaret decides NO WAY as it's raining and she has still not fully recovered from the virus. She waits in the van. I put on my waterproof jacket and off I go. I wasn't really giving much thought to how much effort is needed for six hundred and two steps. I mean try going up twenty steps. I've started, so I'll finish.

It's raining, but it's not long before I wish I had never put my jacket on. I'm sweating buckets. It's too warm for me and too sticky. This is some climb. I can't help but stop every now and then. I keep going and every step I keep saying to myself that it can't be long now, as I literally trudge on. I'm just about to stop again and on the step in front of me someone has put PHEW 301. For heaven's sake, I am only half roads up. This is mental. I think two things:

(1) how on earth did they manage to get their bricks, cement etc., up here.
(2) if I'm only half roads up then if I turn back, when I get to the bottom I could have been to the top.

I keep going. It's forest all around me with strange looking plants and odd looking small birds and I feel like I'm on an Inca trail or something.

One of the things on my bucket list is to visit Machu Picchu. I am well up high and it tells. I think of all the expeditions where you must acclimatise before going too high up. Out of nowhere, suddenly, I'm only feet away from this huge white lighthouse. I got to the top and the first thing I do is sit down. I'm knackered. I get my breath back and my head clears up. I feel back to normal. I think three things this time:

(1) this view is magnificent and I can imagine the beams of light reaching out to sea for miles around and giving all boats a sign of HOPE
(2) the Klaatu song Hope comes to mind from which I often find inspiration
(3) Helicopter. That's the way they must've got the stuff up A HELICOPTER.

Morning Has Broken

PETE:
When I came down the mountain, it was a lot easier than going up. It is very wet and slippery so I hold on to the rail provided. I also get some amazing views and there's a handful of little odd looking birds again following me for a bit. I

also pass a viewpoint with a bench that I never noticed on the way up. Hmmm I wonder how that happened. I finally get to the bottom feeling like an explorer. I come out of the forest and some horses come running over to me.

I get back to the van and tell Margaret all about my excursion and show her some photos. Then we tackle the single track road back to the campsite. Is hard to believe civilisation is only round the corner. It's getting round the corner that's the problem.

We find a spot to park the van so we are closer to see the first beam of light. We find the perfect spot. We set the alarm for an hour before the first sunrise on planet earth. We definitely don't want to miss this. We're going to have another ginger wine hot toddy. Margaret is still shaking off the virus and I am giving her 100% support by having one as well.

The alarm goes off and we're up like shots. We wait and wait and THEN the sky is getting lighter, but no sun yet. As the sky is getting brighter it is getting exciting. The sky goes a mixture of pink, orange, yellow all round and over us. Then, there it is—MORNING HAS BROKEN. What a feeling it makes me feel small in the scale of things. A beam of light covers the field we are in. It's not like getting up and looking out the window. It is a feeling of being alive. You see everything around you spring into life including yourself.

It's at that very moment I say to myself if there is a time to make a wish it's now. The very first moment of a new day. So what did I wish for? Well, let's put it this way. All my family and REAL friends have nothing to worry about. As for all the rest it's been nice knowing you. Bye. With that the first two people to see a new day on Planet Earth. Well, maybe not the first two, but definitely among the first lot. We toast the new day with our ginger wine. So I wonder, will my wish come true?

Hawaii No Wayii

With a new day in front of us, we discover that further down the road is a wee place called—wait for it—Hawaii. This we decide we can't wait to see. If there is another Hawaii, then New Zealand is the right place for it.

It's a very long drive. We've been trying to look for campsites but they are all closed until Spring. All we can do is keep driving to the campsite at Hawaii. We turn on the sat nav it says 6.6 kms and you will arrive at 6. Jeezoh Margaret says I don't like that—what it says 666—I don't like it! When we reach Hawaii it's 6 o'clock. I don't say anything, then Margaret says I knew it! The campsite is next to a graveyard. We're not staying here. I don't care how late it is, with that she does a brilliant back spin and drives off like a New Zealand trucker, mumbling as she goes.

MAGGIE

As we drive down the beautiful coastline, we couldn't find a campsite open since it is the end of Autumn. There seemed to be one open at a place called Hawaii. We decide that we are not that desperate since we notice it is next to a graveyard and the numbers 666 had come up on the sat nav. Time for sharp exit.

This is the third long drive in four days and it is the first time we have done this much in such a short space of time. We decide we need a place to rest and recuperate for a few days. When we come to Opotiki, there is not much going on. People seem very friendly and after a supper of Takihapi fish and chips we find a free camping site at the beach outside town. Just what we need to recharge our batteries.

PETE:

We find another place called Opotiki and as usual with us in New Zealand, it's dark and it's late. We're both very tired. We feel that this is a much better place. The sound of the waves puts us to sleep. While falling asleep, I think to myself. Why would anybody build a campsite next to a graveyard? I promise myself to try and remember I'll google Hawaii—it might mean Devil's Garden or something...

We wake up and although it's a new day, it's not a morning has broken. For a start, it's late morning. That's how tired we are. We've both been feeling unwell, plus there has been a lot of exciting things going on. A lot of driving equals exhaustion. We decide since it's nearly the afternoon we'll just stay the rest of the day and night and do nothing, except walk on the beach. Let me tell you, the beach is lovely. So glad we moved on now.

After a while, what started off as quite stormy afternoon, eases up and the weather changes completely. This is the life man. Love it. I put on some of my up to date insect lotion and go for a walk along the beach. The first thing I notice is all the different types of sea shells.

181

My favourite ones are the ones that look like unicorn's horns. I gather some to give to Margaret. I know this will be a brownie point as she loves unicorns and I gave her one on the Isle of Iona back in Scotland and she lost it.

First though I make a straight line with them, the tallest first (about a finger length long) down to the smallest (a finger nail long). Then I move myself so the sun is behind me and the row of spiral unicorn shells have a beam of light on them. I can just picture it on a calendar.

As we are walking along the clouds are all disappearing although one stubborn one is constantly still. Every now and then I look up at it. The other clouds have definitely moved on. It's still there. Then we see it's an island. Hang on is that a live volcano? We are sure of it.

This is a lovely spot to recuperate. So calm. Just what we need. We head into Opotiki for some groceries and I get my brownie point reward. As we are in town we are going to the launderette and whilst the clothes are washing Margaret asks if I fancy a beer. I make a mental note to gather more unicorn shells.

Staying another night is just the thing to freshen us up. We feel much better. Another day at the beach and that night we look at our photos of 'the first beam of light' and a couple from yesterday. There's a blue orb in one of them and since, for some reason unknown to man, my photo seems to be a live picture, it moves. My own orb. Margaret looks at her photos and she's got three orbs in hers, only hers are yellow.

MAGGIE

We treat ourselves to doing absolutely nothing apart from walking along the beach now and again. The weather had been stormy last night, but by afternoon it cleared. I have always found the sea such a calming place. I love sitting watching and listening to the waves for hours. It gives a feeling of connecting with infinity (and beyond) Somehow, it gives me healing. I feel energised. We decide to stay at this site for another two nights at least.

We see what looks like a volcano in the distance. Just like Iceland this is definitely the land of fire and ice. The two extremes. As it is in Australia, extreme weather is a fact of life here. There is also the constant threat of earthquakes, volcanos and tsunamis. There are notices all over advising what to do in the case of a tsunami.

PETE

We head up the road a bit to Ohope and Whakatane which has an information centre. When we go in for a nosey, there are lots of posters all about volcanoes. We're well chuffed. Our very own volcano and it still smokes. I talk to an older guy who used to fly helicopter rides over the top of it. There's even a laboratory and a few accommodation huts for scientists he tells me. It also can erupt at any time.

I remember the helicopter man saying it takes a good hour to fly there. We are safe enough here I think to myself. I also remember a poster of the volcano with the name of it. I have named it White Island: as such it always appeared to us Captain Cook 1769. Next to it is a Māori interpretation that says the Māori name for the Volcano is Whakatane meaning that which is made visible. I can't help but think of Hawaii and Margaret's 666 thing. Captain Cook went to Hawaii and they liked him that much they ate him. Maybe his bones are in the graveyard. I decide not to mention that to the 'wise woman of the north'.

Whakatane is a cute little place. It used to have a real life friendly dolphin. A little girl and the dolphin were best friends for years. There is a statue of them at the pier.

Hawaii Bay I've learnt doesn't have Captain Cook's bones. Instead of meaning something gruesome Hawaii legend has it that it actually means Place of Gods and could be a reference to Hawai'Loa a legendary figure who is said to have discovered the islands. Mauna Loa and Mauna Kea are the volcanoes. Kamehameha I in 1810 united all the islands into one royal kingdom after years of conflict, so don't say I don't do my homework.

MAGGIE

We head up the road to Ohope to go looking for kiwis as recommended by the girl at the Information Centre at Opotiki.

Ohope has a real feeling of being on the up. The public toilets are great. We got talking to a wee Scottish lady walking her West Highland Terrier. She moved over to New Zealand fifty years ago. Think she enjoyed talking to us. She agreed with us about the crazy impatient drivers in New Zealand and warned us about the truck drivers driving up to the bumper. We told her we had already experienced this first hand.

Be Vewy, Vewy Quiet

PETE:

As we leave Opotiki and drive further up north we see a sign that makes Margaret let out a yee haa. It's one of those kiwi signs and it's in a place with yet another hope in it—Ohope. Well here's hoping we see some kiwis. As luck would have it, when we find a parking place near the beach, there is still some daylight left. So we went exploring.

We came across a sign for a forest walk and as you know I love forests. Maybe I was one of Robin Hood's merry men in a past life or maybe something far different like an Orang-utan. My mum always said I looked like a baby orang-utan when I was born.

We cross the road for a closer look. There is a big boulder with something engraved in it.

Yee haa I hear as Margaret is reading the sign. "Listen to this Pedro," she says, "Kiwis Come out at night. Please be quiet and do not disturb the growth." She then turns round, puts a finger up to her lips and whispers, "Be vewy, vewy quiet."

The boulder along from it has kiwi footprints on it and says, "If you are lucky to see these prints then you just might see a kiwi." Together we say, "If you're vewy vewy quiet."

It also goes on to inform us that you might, in fact, hear them. Right we decide. That's it. We got up at 4.30 in the morning for the turtles so we can go for a walk tonight into the forest.

We go back to the van, get the appropriate footwear and clothes on and of course, a torch. To enhance our chances we check out the internet for a kiwi sound at night. When we hear it, we both decide that they'll be no denying that sound if we do hear it.

We tiptoe in, all excited. The further we go, the muckier it gets and narrower. Then I thought, here we are in a forest at night with just a torch. Then you remember there are no bears or pumas or snakes here to harm you (don't think spiders) All the same, it's great fun and also very exciting.

Every now and again we stop suddenly. What's that? As we've checked out a kiwi's call we count ourselves experts and move on, uphill, relentlessly. I can hear something just above me. There's no monkeys with torches here that I know off. I see lights flashing in a side to side motion and hear people talking. Jeezoh

I think to myself we've no chance of seeing them. They are on the path higher than us and as we pass each other it becomes clear that they're coming down and we're going up. We took the path on the right as we didn't notice the one on the left. It was so dark and my torch was in a straight line. The people we pass haven't seen or heard anything and we tell them neither have we.

On we go, for another good while. We're vewy, vewy quiet. All of a sudden there is the screech we memorised from the internet. We are delighted. Then we hear another screech in the distance in response to the first one. Remember says 'the wise woman of the north' they could be a good distance away. I know I pretend. Walking along I feel great that we have heard them. That means, obviously, they are about.

As we walk on in silence I hear rustling again. I don't say anything to Margaret as I don't want to frighten them. I decide to shine my torch up a little bit higher and further in a straight line. I let out a gasp the torch is shining on a guy's face. Him and his mate are walking through the forest with their lights off. Got to be locals I think to myself. They know what they are doing. Stick to the path, be vewy, vewy quiet and you don't need a torch. As we walk on we realise we are going downhill we are at the other side of the forest and circling towards the beginning again. Ah well, I say at least we heard them.

Walking on there is a little stream running next to the path. It is a quite gentle and soothing sounding in the darkness. I'm in the middle of thinking this is magical when I notice little lights in the branches. Right, I say to myself, Pete calm down they are not fairies or elves as they are not dancing or moving.

I creep slowly towards them and they're still there. They are, in fact, on the banks not the branches and they are on both sides of the river. What on earth is that I wonder? They are actually flickering. I turn off the torch when I pull aside the branches there in front of me is indeed a little magical scene: GLOW-WORMS. There are not many, but enough for me to sit down, here in New Zealand in the silence and in the moonlight looking at a feat of nature. I also realise that it's part of my 'First Beam of Light' wish so just maybe we'll get to see a kiwi somewhere else. I realise we've only two weeks to go. Man! Where does the time go?

We leave the glow-worms and forest behind us glad that we got to hear a couple of kiwis. Quite satisfied with ourselves, we go for a walk before going back to the van. It's still dark and we meet someone coming towards us with one of those lights strapped right on his head. Going in the forest we ask?

Yeah I saw a kiwi here last night. I wasn't quick enough with my camera and the photos didn't really come out. I'm going back in tonight with this one. It's the first time I have seen a kiwi in my fifty five years. This confirms to us that we were in fact very lucky to even hear them. They are even rarer than we thought.

MAGGIE

We thoroughly enjoyed our kiwi walk. We didn't see any, but we heard two calling to each other. We also got to see glow-worms. Nature is magical. Certainly has made our whole trip an exciting adventure as we don't know what we are going to come across next. Seeing creatures in their natural habitat has an uplifting effect. I hope my grandchildren will be able to see these creatures in the years to come. Humans are squeezing all things in their path, forcing them further and further away from their natural surroundings.

Next morning we went to Whakatane Information Centre to find out more about the volcano, which is indeed live. There are boat trips and helicopter trips out, but we decide against it. We spoke to a retired guy who used to fly tourist planes over the volcano. He had great memories and it was very interesting speaking to him. Lots of scientific research has gone on over at the volcano.

I also spoke to an interesting girl from Finland about the Sami culture and compared Sweden and Norway. It reminded me of the cultural differences within the UK. As I am walking away a guy comes back from lunch who originates from Ayr in Scotland. He is a big Ayr United supporter and he was talking about the Auchinleck Talbot result them winning the Junior Cup once again.

As we head on up the Coromandel Peninsula we end up having a scrumptious lemon curd crepe with lemon cheesecake ice cream. HEAVEN on a plate.

We keep on driving to a place Pete feels drawn to. He has mentioned Tauranga a few times and it is the biggest town around this area. This puts me off as I don't like driving busy city roads. I have learned however to listen to Pete's intuition as he is really in tune with the Universe. We head towards a campsite at Cambridge Road Park. I will let Pete explain what happens…

The Wise Elder

PETE:

We leave Ohope and Whakatane and start our long drive to Tauranga. For some reason, I feel drawn to it. God knows why. Maybe it's because it starts with Taur and I'm a T.A.U.R. (US). It's getting late by the time we arrive, but we can still see quite well. I keep looking at all the signs we pass. I love some of the names of some of the towns.

The two names I see on the a sign as we drive into Tauranga nearly makes me faint. There in large letters are four names: one says Otumoetai; one says Waihi: the other two…as large as life say…Bethlehem and Judea. Our friend the Rainforest woman who gave us the Māori blessing had said that she lived in Bethlehem and Judea. That was the second time I thought she was mad. Well, once again she's proved us wrong. We are, in fact, in the region of Bethlehem. Judea is one of the small areas. She had actually lived in Bethlehem and Judea after all.

We find a campsite for the night with the promise that we're going to Bethlehem and Judea in the morning. That night I can't help, but go over the events of meeting this woman. She tells me that she gets bitten by a snake and that I would too be going into the heart of the enemy. So I get bitten by a spider. She gives me a blessing that I can't understand—then a Security Guard says the same words and we realise it is a protection blessing to help with travels and uneasy times. She then tops it off saying she comes from Judea in Bethlehem so now I have come across Judea in Bethlehem in Tauranga. She says remember that Christ did this before me and not to worry. Well, I'm not worrying. It's just all a bit strange. My Lord of the Rings spooky adventure is over. I've recovered from the spider bite. My insect lotion is getting tested and seems to be getting better. I've only two weeks to go. She also said Margaret was a woman of valour. So in that case if Jesus went before me then maybe he can say hello to Margaret for me and she can tell him thanks. I think I am OK now. I feel much better.

In the morning, we head into Bethlehem and to my surprise there are no donkeys or camels. There are plenty of cars and shops and we decide to walk. So this is Bethlehem I think to myself. I wonder where my little friend stayed? Across the road from where we are walking there is a huge Māori Meeting Hall. We would love to sit across from a group of Māoris. We walk up to it.

We know enough of Māori culture now to know that this is a Maura. What we don't know is that we are not allowed in without a Māori blessing or invitation. That is proved when we look in and a group of Māori are sitting on the floor. Before we can do anything, an official looking guy is up and opening the door. Please, you cannot come in here or take photos. We notice a small Information Office across the road as we go in and the young Māori woman, who explains she is a social worker says yes that is correct. You need a Māori Elder to bless you into their Maura. You can however, if you like, come over to the other hall and I'll open it up for you. We are delighted.

It turns out that it's a type of Community Centre, but still feels spiritual with fantastic paintings of Māori culture and whales and the like. There is also a carved figure of a man sitting with his hands, not on his belly, but on his hips and a tattooed face. I bet that was sore, as his face was covered in ink. No wonder he looks angry. Above him, is what I think can only be his name TAMETEAPOKAIWHENUA. Try saying that after a few.

We thank the woman and decide to walk on and not for the first time I hear a chainsaw at work. We walk along a path near a huge lake. It has a jut of land allowing folk to cross onto a small island full of trees. I notice that it must get submerged at certain times of the day, so we don't go. Instead we walk further and a dog comes up to us and as I clap it, a voice says 'here'. I turn round and there is a Māori elder sitting on a tree that has fallen down. We say hello and I ask him what's the name of the tree he is sitting on. An oak he says. It doesn't look much of an oak to me I say. That's because it's not, well not an English oak. This one grows quite fast. So you know about trees he asks. A bit, I say, I'm a herbalist and with that he beckons me to sit beside him.

MAGGIE

Well you just can't make that up. We had no idea that there was a Bethlehem and Judea in New Zealand although our wee tiki woman had lived in New Zealand for a few years. Pete was gobsmacked and so was I. We were going to investigate in the morning.

We realised the next morning that it wasn't one of Pete's dreams. We are actually in Bethlehem and Judea and we just look at each other all day, shake our heads and smile as we wander around exploring.

We end up at Judea Heights and come across a Maura. We get told off as we didn't realise we needed permission to take photos. I totally get it. The need to protect the good energy created in the building. A young social worker takes us over to the other building and lets us see the amazing artwork in the kitchen area.

As we wander further down the path we come across an elderly Māori sitting on a log. He had a chainsaw beside him as he had been cutting lots of logs. Pete bonded straight away with him as it is evident that both of them know a lot about plants, trees and natural medicine. It turned out he is a Māori Elder or Koroua.

PETE

He says, "All this around you. You can eat it all-it's medicine." I pick up piece of plantain and tell him what I would use it for. Yes he said us too. We talk and talk. He has never heard of Echinacea so I find myself telling the echinacea story of Chief Black Elk and the snake. I tell him it increases your white cells and is great for the immune system. He then tells me to do the same thing here he would pick Kawakawa. I am in my glory talking about herbs and swapping stories and names of plants with this wise man.

I tell him we were refused entry into the Maura. He says if you want to wait for a while I can take you to the Maura and give you a blessing as I am an Elder. I can't believe my luck. Here I am talking to a Māori Elder about medicinal plants. Now that really made my day. I ask him about the little island I can see and he says that's a burial ground. Māori's still use it. Even if they die abroad their relatives will, somehow, get their bones over here and onto the burial ground.

We need to move on. I find out his name and he finds out mine. Is that Māori for Pete?

"No," he smiles.

"Remind me of the name of that plant you told me about."

"Kawakawa," he says and to my surprise pulls out a mobile phone and looks up the internet. "We Maoris swear by it. We make tea and drink it from a young age. All Māori drink it he says. It's magical. Here's what it looks like," showing me his phone.

"Now you know what to look for and remember to drink it."

Just as we are about to leave I ask him, "Have you heard of people who have excess trapped energy within their body. What would you give them?"

He got up and walked up to the tree that was on the ground, put his hand under the roots and with a big smile on his face, handed me his chainsaw.

Hot Sands and Mash

MAGGIE

We could have stayed and talked to The Wise Elder all day, but it's time to move on as we want to go to a Polynesian Auckland market on Saturday. First we have the hot springs to visit. As we drive on, we stop at Katikati for a coffee. This is the avocado capital of the world.

We end up camping at a beautiful spot near Taurau Harbour. Beautiful. Pete had found an ale called Deep Creek with a picture of a wolf and the word courage. I asked him if there was a pack of them, no pun intended.

We arrive early at the campsite for the Hot Springs. It's just as well as the low tide is at 11.15 am and the hot water beach can only be accessed two hours either side of the low tide. We are going to dig our own spa pool. We hire a spade, get our 'togs' on and off we go, on what feels like an adventure.

It is nearly winter here, but the sun is still shining. The beach is very busy and everyone seems to be smiling. We can see steam rising. The sand is bubbling and I do a wee dance as it is so hot on the feet. The pool feels blissful. We stay in until the tide comes in. Such fun!

PETE:

I was really happy after my day with such a great character and of course, not many folk get the privilege of sitting on a log with a Māori Elder. He was, in fact, cutting up the fallen tree and all of the handy sized logs he cut were left in a pile for other Māori to help themselves and use them for fuel or whatever. That, of course, explains the chainsaw noise I'd heard.

It's time for a hot bath and a free one at that. We are off to Waikato home of the Hot Springs. Seemingly you take a small spade, find a patch, dig a hole, then sit in it till your bum is too hot, then run the ten feet or so to the sea to cool down. Repeat as necessary. Sounds like fun.

As we travel on our merry way we come across a delightful little place called Taurua. It's a harbour with lots of expensive looking boats everywhere. Quite posh looking. Just like us NOT. It's one of those lovely days again and we have our tea outside. Our luck has really held out weather wise. We have been told a few times that the weather is unusually good for this time of year. Bring it on we say.

We are getting a real dab hand at organising ourselves. We've learned that it is wise to remember that everywhere is usually shut on a Sunday. Mind you with only two weeks to go I suppose we have left it a bit late. We have timed our arrival at the Hot Springs to coincide with the tide as it tells us to do so on the information site.

We arrive quite early, book a site for the van, change into hot springs beachwear, hire a small spade and skip down to the beach. As we reach the beach there are a few folk, mostly women, with jackets and trousers on. Now I think of it the wise woman of the north has her fleece on as well. I think I might not be as organised as I thought. I start to feel a bit naive with my bare chest, shorts, hat and sunglasses. We get to the shore and it is cold. I grab the spade and start digging to keep warm. I remember being told to keep digging until the water starts to bubble up. Then I notice around the holes all the woman have their jackets on.

As I look up a couple start to walk away. In a flash, I nimbly get into their space. Straight away I feel the heat on my toes. I start digging and a flush of hot water gushes up. I shout over 'hurry up and get in'. When she gets in, I hear Margaret say ouch. The water is not warm, it is hot. After a little while, there is enough space and it's deep enough for us both. After ten seconds, I put my elbows in the sand, lift up my bum and legs and try to balance on my elbows. It's then I see someone running into the sea and sitting down as it is getting too hot. I feel like running in as well, but as I think I'll look stupid I decide to just stand up and end up doing a good impression of one of those desert lizards using alternate legs.

I feel sorry for a group of about ten just along from us. Every one of them has a spade. They all dig together and then they all sit down in their pool together, only for a wave to come up and cover them. They are evidently not as organised as us and the tide has just come in.

MAGGIE:
By the time we walk back, we are hungry. Time to try out the mash. I have not had instant mash in a long, long time. We got some bratwurst from the butchers and a tin of beans. Bangers and mash with lashings of butter and a dash of milk. Fast food, but it was a one off and delicious when you're hungry. After spending ages doing the laundry, it's an early night. In the morning, we are heading to a Polynesian Market (Fleah) near Auckland.

We arrive at the market around 10.45 am. The car park is mayhem. Not what I expected. It is mainly Asian stalls with only two or three Polynesian stalls selling the same garlands of flowers, tee shirts and hats. We did manage to find a stall supposedly selling Hangi food which is a traditional Māori method of cooking food. It uses heated rocks buried in a pit oven (umu) and our Lovely Wedding Celebrant recommended that we try one if possible. It consisted of potatoes, cabbage, pumpkin, stuffing, chicken and pork. Was interesting to try it.

We wandered around and it didn't feel that safe. It seemed like a poor area and a bit run down. We went into a shop and asked what the ball like fritters were. Turns out they were banana doughnuts—five for a dollar. Really enjoyed them, but a bit like churros I wouldn't eat them very often.

Then it was time to find a park and ride car park to allow us to take the bus into Auckland. Went on a bit of a wild goose chase because the term park and ride is not used in New Zealand. Finally we end up at Albany Bus Station which is a north suburb of Auckland and which has 24/7 parking.

As we jump on a bus heading to the city I notice that we need to pay cash. I just assumed I could pay by card I had not had a chance to go to an ATM. I quickly say adult and senior and it comes to $11.50 and I have $11.60 dollars. That was very lucky. I feel stressed with all the driving about. Pete was full of praise. I realise: YES, I DID WELL and gave myself a pat on the back.

PETE:

Back up at the van we look for something nice and quick to make. I look in the fridge and there are sausages I had forgotten about. Margaret rummages in a box. Beans she shouts and pulling out a packet of mashed potato I say nice one! I remember picking it up when we first arrived in New Zealand and thinking it would come in handy as you just add water. Bangers and mash are delicious when hungry.

Next day we go to a Polynesian market looking for kawakawa. No-one has heard of it. We are not that impressed so we don't stay too long. We will try somewhere else once we get to Auckland.

We chance upon a place called Albany and there is a huge 24/7 car park. Since it is open all day, seven days a week and to save any more stress and turmoil it is ideal. Driving in Auckland, which is vast and very, very busy, is hectic. We park the van and catch a bus. It's a great run, the buses have a wide lane all to themselves running along the coast. No cars, lorries or bikes allowed. Apart from the bus stops, it's a straight run through and hassle free. We sit back and enjoy the ride. Once again it is a lovely day.

Auckland for a Day

MAGGIE:

We have made it our mission in Auckland to find some kawakawa tea. After trying a few places, Pete spots a place across from us and lo and behold they have it! The lady we speak to seems very knowledgeable and it turns out that her son is a Herbalist and she is very interested in Pete's story. In fact, she would love Pete to meet her son (he is there every Monday) and invited us to come back. She kept Pete's flower power card. We walk out the shop with our kawakawa tea and homeopathic seasickness tablets and the first thing we see is a sign saying Vogel. The Herbalist Alfred Vogel happens to be an inspiration to Pete.

It is a day full of signs and synchronicity. We already knew this, but because we have slowed down, given ourselves time to go with the flow, we have tuned in. The thing is that really every day could be like that for all of us. All we require is to keep calm, pay attention and listen to our hearts. Help is always there.

As the day goes on we see the name Vogel in numerous places and also images of wolves and to top it all the Royal Wolf Poster saying you can do anything with a Royal Wolf. We go to a Brewery Bar and end up sampling a couple of real ales and I get talking to the girls behind the bar. It turns out that Deep Creek Brewery is very near to where we have parked our van in Albany.

We are getting the bus back to Albany and I notice a twinkly faced woman waiting at the bus stop. The bus comes and we go up to the top deck to try and get a seat at the front. There she is, sitting at the front. She seems very, very pleased to see us as if she was expecting us. She gives us a radiant smile and compliments me on my hat and 'Sweet As' badge.

We get talking and she tells us she is originally from Texas. When she hears of our travels, she says she wishes she had someone to do that with. Not necessarily romantic, but just company. She says she is quite lonely and we have to GO FOR IT while we can. While the Earth is still in one piece. She talks politics and tells us she has been to a Writers Festival. She shows me the Programme and makes a point of giving it to me saying she wants me to have it.

Pete whispers that she could be the sister of the wee healer woman from Australia who gave us the blessing. Just at that a bell sounded. I thought of the film It's a Wonderful Life and Clarence.

PETE:

In Auckland, we decided to look for place that sells Kawakawa tea. It seems hard to find. Finally, I see a shop and can tell by the goods displayed in the window that it is a herbal shop. So it's worth a try. The woman has indeed got Kawakawa tea and after buying some, I notice tinctures on a shelf labelled with their Latin names.

I ask to speak to the Herbalist in the hope that he can tell me what he prescribes Kawakawa for. The woman in the shop tells us that the Herbalist is her son and he is not in until Monday. She seems interested in what I say and takes one of my cards for her son. We thank her and walk out the shop and the first thing I see is a huge van with the word Vogel on the side. That is the name of the Swiss company I did my training with. Then I see Vogel written all over the place. Even a street is called Vogel.

Maybe it's a sign that it is time to crack the recipe for a real insect lotion using all the herbal ingredients I've been experimenting and using so far. I know by using it that I am getting closer. I just know that there is only one missing ingredient that will make it do the trick.

We have a nice day in Auckland and the bus we get back is quite late. We head upstairs hoping for a front seat. There is a woman sitting there, so we sit behind her. Hello she says my name is Ann. Ann with an e. She has the same kind of attitude as our old friend The Rainforest Woman from Australia. As I sit back, Anne and Margaret get talking. I don't know what they are talking about as I am too busy looking out the window.

MAGGIE:

We wake up and as we want to visit the Deepcreek Brewery when it is open tomorrow, we decide to hang around the area. We visit nearby Silverdale Shopping Centre and Pete sees a cafe called Silver Moon. It is an absolute treat.

We then wander the shops. I go in and get advice from a mobile phone shop. I say to the guy we are a bit techie ignorant and his eyes roll back up to the top of his head and sighs. You are not alone. Keeps me in a job lol.

We find a campsite nearby at Red Beach and nothing much happens. We enjoy the simple things in life, so we go for a long walk. Next morning we awake and head for Deepcreek Brewery. First stop is the Op Shops of which there are three. Pete finds a tea strainer for the kawakawa tea and the lady asks for 3 dollars and he nearly faints. Bear in mind that he got genuine Gortex walking boots for three dollars earlier in our trip. Can't win them all lol. It is for a good cause.

Our visit to the Brewery is fun. The lady serving told us that they had just been over in Australia at an Awards Ceremony and they won Brewery of the year. We are not surprised as the beer is tip top. We love the double hops. We tasted a few and opted for two litres of Lipus. I asked her what the name meant and she said lipus oil is a cousin of marijuana.

The shop had some tee shirts for sale and Pete fancies getting the Deep Creek wolf but they don't have his size. She was ordering more so we said we would probably call in on the way back down the road. We can have one for the road as well before we fly home.

As we head up the road I took a turn for the Kaura coast which is more off the beaten track. We are in search of the ancient Kaura trees.

PETE:

In the morning, we head first of all to Silverdale and I spot a cafe called 'Silver Moon-best coffee in town' and they are not wrong. After breakfast, we explore the shops.

We are heading to the New Zealand Brewery that makes my favourite ale of the moment-Deep Creek—Wolf. The woman behind the counter explains that the hop they use for our favourite one was in fact a relative of the marijuana plant. Honestly, we didn't know that. We get a couple of litres anyway and tell her we will be back as I like the look of the Wolf Creek T-shirt on sale and they'll have my size in a week. Perfect!

We head up the road and decide to go the coast road which should be interesting. First, we are looking for a library to do some emails. We go through a few small towns with nothing much going on and eventually come to a place called Dargville. There is a library.

I see a logo for the council and it has the same shape as the one in my dream when my friend Gabie came to me. I asked the librarian what the logo shape was. She explained that it seems to be a Koru which is some sort of Māori spiritual emblem and if you wait for my colleague, who is Māori, he may be able to tell you more. Hmmm this is going to be interesting because this is the only time I have seen the exact same shape which was in my dream.

We hang around and when her colleague comes in, he is definitely Māori. He is smiling and huge. He tells us that Korus are very, very powerful protectors and to choose one carefully as it concerns the past as well as the present and the future. The shape is very important to the individual and represents health, well-being, protection and prosperity.

I tell him about the shape of mine and he says did you make this shape up? No I dreamt it I tell him. "Even better !" he says with a huge smile. He says this is mine as he puts his hand under his shirt and pulls out a huge piece of pure green jade. It must have weighed a ton round his neck and was about the length of his arm.

"It never leaves my neck and I wear it with pride. It represents all of the strength my ancestors and my own strength in present life." We thank him. As I leave I think why can't we all be like that and I wish I had one!

MAGGIE:

We drive and drive looking for a library and also a place to have a picnic, but to no avail. Eventually we come to a town called Dargville right on the edge of nowhere and there is a library.

Pete spots a Māori symbol which is exactly like the one in his dream. It turns out to be similar to the logo for the local council. When we ask, the helpful librarian reckons it is a Koru, but can't be too sure. She invites us to wait until 4.55 as her colleague the Iwi Liaison Officer, is attending a meeting. She is sure he will be able to help. Perfect! I reckon with that job title he must be Liaising between Māori culture and 'white' culture.

Sure enough, in comes a big fellow with a big smile and a big heart. I think he is happy that we are asking him a question about his culture. He explains the history of the Koru symbolic of the silver fern frond shape. It symbolises new beginnings and infinity—past, present and future.

Ka hinga atu he tete-kura—ka hara-mai he tete-kura.

As one fern frond dies—one is born to take its place.

We reckon that was what the dream was about. Gabie was telling Pete it was a new beginning for him and he was protected by his loved ones and his ancestors. When Pete said he saw the shape in a dream, our friend didn't bat an eyelid. He just smiled his big smile and said, "Even better!" Most interesting and also most helpful in interpreting Pete's dream. Thank you.

PETE:

I am going to keep my eyes open everywhere we go from now on to see if I can find a piece of jade. I'll be looking tomorrow as we are heading up to the Waipoua Forest where they giant kaura trees are.

After doing our homework, we learn that there are two gigantic kaura trees from ancient times still standing and producing leaves. They are the two oldest trees in New Zealand. Very delicate and very fragile. One has a 16.5 metre girth that's about 54 feet and he is about thirty metres high—98 feet. He is called Te Matua Nhahere—Father of the Forest. The other is 13.7 metres girth that's about 45 feet and 51 metres or 167 feet high and he is called Tane Mahuta—Lord of the Forest. Both are over two thousand years old and still living.

I mean two thousand years old—that is older than Christ. Looking forward to seeing these guys.

The Two Towers

MAGGIE:

We arrived at the rainforest late, as usual. We manage to find a space in the dim light which we reckon will be OK. We talk to the Night Warden and he says, "Kiwis? you'll be lucky, you might hear them but, they are getting rarer than hen's teeth." Oh well keep the faith babe. We didn't see a duck billed platypus or an Aborigine playing didgeridoo till the last day in Burleigh Heads, just before heading to New Zealand so who knows?

PETE:

We arrived at Waipoura Forest campsite just as it was getting dark. We find out that we can pay at the cafe in the morning. We go straight to bed. After a coffee in the morning, as we are getting in the van a young guy comes up. He says, "Excuse me are you the owner? Can I get in the cafe?" I tell him I am not but that the owner has just left. Oh no he says my friend is in the toilet. I can hear him. He is locked in. I can't help but laugh to myself. I've waited years for this one to happen. Eventually, the owner came back and all ended well.

When we got to the forest, we find it is the real deal. Before entering the forest, we have to walk over three iron grids that spray different kinds of disinfectant on the soles of your boots. I think back to them doing a boot check at the airport when we arrived. They are very concerned about the flora and fauna in New Zealand. They don't want any foreign seeds brought in on the soles of boots. As we walked through the disinfectant chamber and into the forest I can't help but think of the fight against the 1080 poison that is killing paradise.

MAGGIE:

We headed to the trees just after going to the cafe. We left just as the manager was locking up. As we leave the manager jumps in his 4 x 4 and drives away. It turned out he locked a guy in the toilet by mistake. All ended well as we saw the guy later on. He was on the same path going to see the sacred trees. The tree is awesome. There is a Christ like quality about it with its huge branches like arms stretching up towards Heaven. Huge. I can't take my eyes off it. I feel truly humbled to be in its presence. Felt it's spirit. It's patient wisdom. We can both see a face on the trunk and wonder if anyone else has seen it. Probably, if they took the time to look.

We speak to someone at the entrance who tells us about the blight. We had to clean our shoes with disinfectant before walking in the forest and are told we must keep to the path. The ancient trees are host to over a hundred different plants. If they feel they don't want something, which may invasive or dangerous, they expel their bark. Just like humans they have to heal from the inside out.

PETE:

It's a nice walk and being a rainforest it obliges with rain. Some of the trees are stunning and look as if they have got faces. They would be perfect as extras for the Tree Ents in Lord of the Rings. There are huge ones who look like Treebeard.

We move on, as according to the sign posts these big trees are not the ones we are looking for. There are birds in here as well. We have to keep to the walkways and some of it is raised to protect the roots as their roots are near the top of the ground and spread out. They're also hollow and brittle. Walking on them can kill these gentle giants. Every time we see a larger tree we think this could be it, but the signposts tell us to keep on walking.

The Information Boards tell us that there is a fungus which is threatening to kill these trees from the inside out. It's hard to believe that trees this size can be brought down by a fungus. As we walk on deep into the forest we come across a sign Te-Matua-Ngahere twenty metres to the left. As we walk down I think I see part of a cliff between the gaps of the Kaura trees. Further down, it looms right in front of me. There he is: The Father of the Forest. I have never seen a tree this size in my life. It's indescribable. This makes all the other huge trees round about look tiny beside it. Awesome.

We walk back up looking at all the trees all around us. As we walk we notice a crowd of people coming down the path and one of them is the boy in the toilet.

We disinfect our feet again and head back to our trusty steed. We get out the trusty map and discover that there is a place not that far away from here called Kawakawa the same name as our Wise Elder's herbal tea. We decide we need to go, but as usual, it's getting dark. We eventually find a place for the night: a campsite on a farm. It is ideal, the two dogs are both big softies, the woman is friendly. It's busy. Plus she tells us take as long as you want in the morning to get ready to leave. I told you she was nice.

MAGGIE:

We head on and come to Hokianga Harbour and can see huge sand dunes across the stretch of water. We go into the Information Office and the lady there tells us about the 90 mile beach and the huge sand dunes there. You can actually board down them. It's a long bus trip and we decide it is not for us this time. We are happy to look at the dunes just across the water.

We have noticed a town called Kawakawa on the map. We feel it is a must to visit due to our talk with the Wise Elder. It is getting dark and we stop at a town called Kaiwoke. We get an uneasy feeling there. We decide not to hang around. We see a sign for a farm campsite called the Cow Shed. It feels safe here.

The shower in the Cow Shed is fabulous and we don't have to hurry. We decide to head for Kawakawa via the local hot springs. It is five dollars entry and looks basic but intriguing. There are some tanks and some pools. We saw a temperature saying they were 52 degrees and we could see the water bubbling. Eek. Time to move on.

Kawakawa looks like a nice town with a steam railway and famous toilets by a world renowned Austrian artist called Freidensreich Hundertwasser. We go to the Railway Cafe for lunch. The cheesecake looked scrumptious but I opted for a healthy smoked salmon salad. The girl asked how our day was going. I said great but wondered why so many police were on the road. They are heading to Kaiwoke. It's a ghetto. So our gut feelings last night were right. So glad we listened to them and headed out of there as quickly as we could.

Three Hundred Dollar Sandals

PETE:

On our road to Kawakawa, we pass a sign saying Hot Springs. We stop to investigate and find that the hot springs are surrounded by wooden sides making them look like big hot tubs. We notice the Information Boards it says admission five dollars and then we see that each tub temperature seems to get warmer and warmer and one is much hotter in fact 52 degrees centigrade. Body heat is 37 degrees so that's far too warm for us. Nice thought, but no thanks.

With that it's on to Kawakawa and I'm intrigued that it has the same name as the herbal tea. We end up in a town called Kaiwoke but quickly move on as it just didn't feel right. My honed up spider senses were proved right the next day in Kawakawa when we found out that the police cars and ambulance with blue flashing lights were heading to Kaiwoke. A girl we were talking to said you were so right to move on. It's a ghetto full of trouble.

We scout around looking for kawakawa tea and can't seem to find it anywhere. Then we notice an op shop my favourite charity shop so in I go and it's not long until I find a pair of real leather sandals not unlike the one's the Wise Elder was wearing. I pick them up and they look as if they will fit. I try them on. Perfect. I turn them over and the sticker says three dollars. Even more perfect. I show Margaret them and she thinks it's a bargain.

I take them up and the Māori woman looks at them and then at me and says that will be three hundred dollars please. CASH. We all start to laugh and she says I don't know who put the sticker on but it says three dollars so three dollars it is. Mister you got a real bargain there! Aye I replied and she said that is Scottish if ever I heard it. Och Aye I said in a deep voiced send up.

As we are talking away I tell her about The Wise Elder and the Kawakawa tea he told me about. I comment on how the town is called by the same name. That is because there is plenty of it about she says. In fact, she says come back at 4 o'clock, If you follow me in your van I will show you what it looks like in the wild. Will save you buying it and it will be fresh. By the way, if it is full of holes as if the insects have eaten it then that's the best ones. The insects seem to know what leaves have the strongest and best ingredients. I thank her. I will keep that in mind when I pick any herb now.

At 4pm, the kind lady takes us out of town and shows us what the plant looks like. When she takes us back, she meets a friend and introduces us. Hey she says you could have saved yourself a drive there's some over there. I pick it and use it in my bath she says and I remember The Wise Elder telling me exactly that same thing.

I go 'over there' to investigate. I pass a public toilet. It is the coolest toilet you will ever see. A famous guy, an Austrian, built it out of broken glass and ceramics. It's stunning and colourful. There's another famous one in Vienna. There is even a roof garden on them.

We see the shop with two pillars made out of the same, so we go over to look. It was definitely designed by the same guy. We found out he stayed here and even had this place at one time. We go into the shop and lo and behold guess what is there? Lots of Koru made of jade greenstone. "Maybe you'll see your design from the dream in here Pete," says Margaret ever the optimist.

"I doubt it I think it seems to be an unusual one. When we walk around and look at the designs, alas I don't see mine. It's like two curly letter S facing each other, but barely touching each other," I say. A good ten minutes goes past and Margaret says, "Excuse me can you please open this cabinet, please." She holds up a Koru, "Look at this Pete, what do you think?" When I look it is my Koru. It's almost the exact shape. The woman says, "It's unusual this one, not like the other designs. In fact, it's the only one I have seen like that." "That's it then," Margaret says, "it's yours. You've got to get it. You might never see anything like that one again. It's a present. You can't buy your own anyway. It has to be gifted. This is my gift to you."

I walk out of the shop with my own Koru dream design, bought for me as Māori tradition implies. Just like our Māori friends I wear it with pride and it never comes off. Thanks Margaret. I love this woman to bits xxx

MAGGIE:
We go into Charity Shop to see what bargain Pete can find. A pair of leather sandals for three dollars. I paid eighty pounds back home for a similar pair I'm wearing. The lady looked twice and said that will be three hundred dollars cash please lol. You got a bargain. We got talking about Kawakawa and she insisted

that she was going to find a plant/vine to show us. She took us into the gift shop next door as she sold kawakawa tonic/tea/balm made by a local woman. She said to meet her at the shop at 4pm when she finished. Then she went back to work.

We saw the greenstone in the shop and after looking for some time, lo and behold, I saw the shape from Pete's dream. It's the only one like that we have seen on our travels. The entwined Double Koru is a Māori symbol that is said to represent two lives growing together. The Sky Father and Earth Mother joined. I think that it is very appropriate for us. It really does feel that we are growing together. He wears it with pride. Pete's dreams never fail to amaze me.

When we meet up with our new friend and follow her in the van, we are surprised to see the plant is like vine leaves and much bigger than we thought it would be. This is so kind. She gave us lots of great information about the plant and about REWA a reconstructed Māori village near Kerikeri. As a thank you, Pete gave her some mullein oil he made in Australia, (good for the ears) What a lovely lady. She recommended the Waitangi campsite and that is where we ended up.

Waitangi is famous for being the place where the Treaty between white settlers and Māori chiefs was signed. We decided against going to the Treaty grounds as we felt it would be quite sad.

Aladdin

PETE:

We leave the joys of Kawakawa and Waitangi and drive to Kerikeri. It's a right busy town to stumble upon this far north. In fact, we both end up getting haircuts. I get a beard trim and Margaret gets her eyebrows threaded. I think she forgot she was in New Zealand lol. It was one of those sore looking ones where you lie back and somebody gets a piece of thread and swipes it across your eyebrow taking all the hairs with it. "Ouch."

Later, as we walk about and we're heading to Noah the van we spot a picture house. We go in to find out what's on. "Nothing much on except Aladdin and I've seen it a hundred times," I say. "What? With Will Smith?"

"Eh? Will Smith? Don't you mean Robin Williams?"

"No," she says Will Smith in the new one. A new one with Will Smith— never heard of it. We decide to go in and it is great. Much, much better than I expected. To top it off, when the film finished all the youngsters got up and danced at the front of the stage to the theme tune as the credits rolled.

We found a charity shop in Kerikeri and yep, I went in. This one was huge so my senses were well pricked. I see a book, I recognise it straight away. Lord of the Rings Official Film and Locations Book. I pick it up and buy it without looking at it. I'm feeling quite chuffed. When I read it in the van, it has many of the locations we have visited with lots of pictures. A great reminder of our trip.

MAGGIE:

We are heading up to Kerikeri. Really nice town with laid back vibe. We both got haircuts and I got my eyebrows threaded for the first time. Ouch. I am coming home with my natural grey hair. I like the colour. It's nicer than I thought it would be and it's hassle free.

There were lots of school children out with placards protesting about Global Warming and we could see they really meant what they were saying. We have certainly seen first-hand how much Australia and New Zealand are suffering from global warming.

We decided to go to the pictures (cinema). Have only watched one film and no TV for the past three and a half months. Have not missed the TV and it's subliminal brainwashing one bit. The picture house was cosy with comfy seats up the back. Aladdin was really, really enjoyable. Will Smith made a great genie. Both leads were adorable and the songs are fab. Disney at its best. When Aladdin put his hand out to Jasmine and said trust me, Pete put his hand out to me and said trust me. It meant a lot. I trust him with all of my heart.

The young people there were very respectful and afterwards one young teenager came up and asked us if we enjoyed the film and then proceeded to give us his critique. He was a young film critic in the making and gave it the thumbs up. A lot of youngsters got up at the end credits dancing to the theme tune. Joyful :)

PETE:

We camped at Paiha, a campsite on the outskirts of a forest. It's a lovely place and quite busy. We have a look at the forest and there is another one of those ramps to walk across and disinfect. We are going to reconstruction of an ancient Māori village and we don't want to be late so we are up early and move on. When we reach Rewa Village area, it's up on a hill, where they would have

203

had a vantage point way back in the days of marauding tribes. You have to walk to get there. First we walk in the park next to it. It is a real treat with green grassy slopes, lots and lots of white birds, cute little boats and a path leading into a nice meadow.

The gate into Rewa is small, wooden and white held up by two white fence posts and it swings on hinges when you open it. The odd thing is there is no actual fence. You can, if you wish, just walk past it on either side. For some strange reason, we go through the gate and I bet most people who come here walk through the gate just because it is there.

We know we have reached the entrance to the village when we turn to go up the hill there is a large wooden archway with dozens of carved, angry looking tiki Māori faces staring at us. As you go under the arches there are wooden steps leading up to a sign that says, Rewa's Village—a replica of an ancient fishing village and native plant garden. That's the very thing for me.

We walk round about. We pass huts, thatched stores and a communal hut where everyone would sit round the fire. I decide to go into the garden and in front of me are loads of wooden tags with numbers on them. E1, A1, A2, B4 etc., which explains a dream I had a couple of nights ago. I dreamt that I was looking at a load of numbers. Here I am wandering round the native garden looking at the numbered labelled plants and their medicinal properties. There it is: Kawakawa and the leaves are full of holes. I can't resist a photo next to it, even though I know in the back of my mind that where there are holes in leaves there are insects. According to the notice Kawakawa is also known as the pepper tree.

There are loads of things here and I walk through the mangroves just so I can tell people I have actually walked through a mangrove. Margaret has brought an information sheet with a list of the names and uses of the plants next to the appropriate numbers. She says she will get it laminated for me. Bless her. I want to get her something. When it pops up, I will know.

As we drive back to Kerikeri I think to myself that was good and not a bite in sight for once.

MAGGIE:

We go to a campsite at an ancient forest. We arrive at 8pm and leave at 8 am. It is a beautiful morning and we want to make the most of it. It's great to be up early and everyone we meet is saying how good the weather is for this time of year. Winter officially starts in seven days' time, the first day of June.

We are going to visit Rewa a reconstruction of a Māori settlement, but first of all we walk around the old schoolhouse, the park and visit the Church where the first Christian mission was built (rebuilt in the 1970s). It is named after Saint James who walked the pilgrimage Santiago de Compostela from France to Spain. This is a walk Pete and me would like to at least, partly try one day.

Felt very humbled by the stain glass window where two Anzacs stand. It says: We gave our today for your tomorrow. As I reflected on that, I also reflected what Jesus did for us and said a prayer of gratitude to all who have died so we can be free.

Rewa was very interesting especially since the girl at the door was so friendly and informative. We find out that kawa and kawakawa are absolutely different meanings. Kawa is Māori protocol and etiquette and Kawakawa is a small shrub. She said that most Maoris now have lost much of their knowledge and use of ancient customs.

Pete found the native garden very interesting with their names and properties listed and numbered. Pete's dream from a couple of nights ago now makes sense. He must have known he was coming here in his dream time. Heard about how kawakawa was so important in Māori life in ancient times.

When we got back to the van, someone had stuck a tiny globe to the door handle (which we still have in the front of our camper van) We are definitely going back to Kawakawa on the way back down the road before we head to Auckland for Friday's flight back home.

PETE:
Back at Kerikeri we pass playing fields and a game of rugby. We feel we have to stop as New Zealanders are so passionate about rugby. We hope to see them do the Haka. We reach the field and miss the Haka, but get to see them in a huddle. Then the game starts and it's not long until the referee blows his whistle and somebody is holding their face. He gets up and gets on with it. What seems like a couple of seconds later his whistle goes again. Another guy is holding his leg. There's loads of scrums, ball throwing and knocking the stuffing out of each other.

The scoreboard across from us said 20-17 to the away team, when all of a sudden the two young boys keeping the score put up 27-17. The teams are still playing and Kerikeri are getting beaten. When the referee blows the final whistle, both teams limp off. Some holding their shoulders, their arms and bandaged heads. As we go away Margaret points out the Referees top says sponsored by a firm of funeral directors.

We head back to the centre of town and I manage to get Margaret a nice big Swiss chocolate ice cream. It was delish. Happy as bunnies we head back to Kawakawa. Speaking to a guy there who says he is happy that Kerikeri got beat 27-17 as they are toffs. With that we go to our beds wondering what tomorrow will bring.

MAGGIE:

As we drove we came across a rugby club and a match was playing so we decided to stop and watch New Zealand's favourite pastime. A real life rugby game. Not for the faint hearted and yet played in good spirit. After the game, about six guys from the same team came off limping, arm in sling, leg in straps, walking with the help of team mates, looking weary. They were beaten in more ways than one. Thought it was appropriate that the sponsor was a firm of funeral directors.

Pete bought me a dark chocolate and cappuccino Swiss ice cream and it was delicious. RED SQUIRREL! (our code for a guilt free purchase)

At Kawakawa, a guy spoke to us and when he heard Kerikeri got beat at the rugby he said great coz they are a bunch of toffs. Reminds us of home a bit. In fact, the Māori use the word Aye for yes so reckon we are kindred spirits.

Through the night I felt that heat coming from Pete's body. I didn't realise he had been bitten again until the morning when he turned round. His face was all swollen, his eye and nose puffed out. It's Sunday. No pharmacies on a Sunday I realised. I need to take him to Hospital. Pete refused to go.

PETE:

I woke up in the morning and felt like shit…little did I know that I looked even worse. My eyes were swollen so much that I could hardly see and my nose was swollen up as well. I can only think of the garden in the Rewa Māori village. For God's sake, the last time I got bitten my whole, I mean my whole, body suffered. Now it's the turn of my eyes…and my back was itchy again.

Margaret suggested the Hospital but I am not keen. We pass a Church and as it is our last Sunday in New Zealand, Margaret decides to go in and investigate. No thanks I say, you go. Margaret goes away in and I am sitting in the front of the van. Suddenly I feel the need for some fresh air. I open the door wide. Holy shit the last time I felt like this was years and years ago when I had a 'whitey' with too much marijuana. I feel I need to rest: I wake up and hear birds singing, the wind blowing. In fact, I can hear very clearly and my vision is clear as well. The only thing is that I'm lying on the Church car park ground. I get up no problem. I feel much better as if the poison in my body has left. As if a curse or something has been broken. Although I am still reluctant to look in the mirror.

When Margaret comes back from the Church, she's got lots of cakes and stuff. You'd get a piece at anybody's door I say. She says you are a bit better by the sound of it. Still, I think that you need to see a doctor. This time I agree, but only because I have ran out of kawakawa tea and antihistamines.

The Doctor we see is a Māori and she is a gem. I take off my sunglasses and she takes a look. I could shake Margaret when she says, "Pete collapsed this morning." I'm fine, honest, I protest.

The Doctor asks a few questions then smiles, looks and says you seem OK now apart from you need a good antihistamine. Thank God for that. I like the sound of good antihistamines. When all is done, I think the Doctor has taken a shine to us and she starts telling us about the fantastic youth hostels in New Zealand and wait for it: kawakawa tea. To be honest it is not too long before I start to feel better. Whatever was in the bite seems to have gone now.

We do a last bit of shopping in Kawakawa. In one shop, a record comes on and again it's as if it has been waiting for us. Black Velvet. We take that as a sign and buy ourselves the ingredients for a Black Velvet—Guinness and Prosecco. It's time to celebrate after all. I've no more itch and we only have four days left in New Zealand.

We managed to make the Black Velvet, which is always nice. You have got to do it right or it's a waste of Guinness and Prosecco. You pour half a glass of the Guinness first. Let it settle, then you fill it up with the Prosecco, using a spoon. If you don't pour it very gently over the back of the tablespoon, it won't flow evenly on top of the Guinness and it will look just like a pint or half pint, BUT if you do it correctly, then patience is rewarded. The bottom half is black the top half is clear. It best to drink it without mixing it if you can. You taste the sparkly wine and then the Guinness comes through. Warning when you sit the glass down the force of the glass makes it mix. So…it is up to you, but, we usually just neck the glass to get the proper taste. Much better than wasting it.

MAGGIE:

I decided to go to a Church Service in Kawakawa since it was our last Sunday. Pete didn't want to go. The service was 9.30 till 11am including tea. The congregation were all elderly and it was old fashioned, but it was sincere and heartfelt. The minister was a lovely lady and the sermon felt right. That Christ is in all of us across the world. The prayers were for peace in the world and in all of our hearts.

Communion also took place and for the very first time in my life, I really 'got' what communion was all about. I felt such 100 per cent gratitude in my heart towards Jesus for what he had done by leaving his pure love energy here on Earth, for us. At the end, we all had an opportunity to say our own private prayer into ourselves. I prayed for Pete to get better. The service finished with Sing Hosana which was very upbeat. One member said we should sing more hymns like that and I agree wholeheartedly. I get invited to stay for tea.

Once in the hall there was a lovely spread of food. I had mussel patties for the first time and very nice homemade pancakes. I got talking to two different ladies who had been to Shetland on different occasions. One summer—one winter. Both loved it. They felt it had a strong economy, beautiful houses. Saw Orcas. I mentioned that we were hoping to go too this Summer. Someone had stuck that wee globe on our van and I took that as a good sign. More travels.:)

The lady Minister came to speak to me and was saying how expensive New Zealand was for food considering how many resources and products they have such as butter, honey and wine. I said it reminded me of Scotland paying big prices for our own products such as oil, whiskey and honey. As I left she gave me a pile of pancakes and mussel patties for the road. Thank you for your sermon and your kindness.

When I went out, Pete looked very pale. He said he had collapsed in the car park. He felt something come over him as if something was leaving him and he was left with a clarity of senses. As if he could hear and see for miles. Never thought of the coincidence until we were talking later. I told him I had a very deep feeling of love in my heart at the communion. I felt the sincerity of the small congregation and the love of Jesus in my heart when I prayed for Pete. Pete reckoned that it was at the time of the prayer that he felt something come over him. The feeling of illness had left him. I felt he still should to get checked by a doctor at the local hospital since it was Sunday. He agreed to go.

The doctor was a real character. She listened carefully to what Pete told her. She prescribed antibiotics and antihistamines. She took a shine to us and gave us advice about Youth Hostels and Kawakawa. I wish our doctors had her sense of humour and joie de vivre. That in itself is a tonic. Thank you.

PETE:

We end up at a place called Whangarei. It's a nice place. The cleaner is a nice guy and we ended up talking straight away. We end up giving him our two wine glasses as we won't need them now.

He says, "Follow me" and shows me an old bus and tells me, "That's my home, has been near on twenty years." As I talk to him he says that he has been all over New Zealand when he first got his bus. Sure enough it doesn't matter where I mention (and we have been to a lot of remote places) he has been. I decide to mention my spider bite and ask him just how rare this white tailed spider is. "Oh yeah," he says a few folk have died with that bite "man you are lucky dude" I can't believe my ears "You know what. I thought I was going to" Well it's not only them that can kill you there's the daddy long legs. Now I'm worried there are millions of them! Yep but the good news is that they might be toxic but their mouths are that small that they can't penetrate the skin. I can tell by his face he believes what he is saying. After a good old chinwag, we say cheerio and clink the glasses. For a while, all I can hear is Man you're a lucky dude.

MAGGIE:

Time to head to Whangarei (Anne with an e mentioned this town to us on the bus. It was as if she knew it would be important to us). We arrive on a Sunday and everything is closed. It looks a bit down on its luck. Black velvet had come on the radio we take it as a sign and get the glasses out. We woke up to rain but once again it cleared away and it ended up a beautiful day.

We head up the town 'basin' to find some shops. We found a beautiful cafe aptly named Treat. Could tell and taste that every cake and pastry had been made with love. We sneaked a miniature Baileys into our black coffee. Double treat.

We ended up in a couple of shops where the staff were lovely. The first lady was Canadian and she had Scottish ancestors. Then in the next shop the New Zealand lady was married to a Scot and did a great accent. "There's no such thing as bad weather just the wrong clothes." TURTELL and not turle lol. She guessed that Pete had been a ginger baby as all fair skinned blond/ginger had sensitive skin. Pete told her the orangutan story. She said only a mother could get away with saying that.

Then we headed for the van and realised four hours had passed without realising it. We came across another interesting Hundervasser shed called 'the seed'. It's time to head for the Bird Rescue Centre and we didn't realise just what other treat we had in store.

Sparky and the Glow-Worms

PETE:

As we arrive in the town centre, we notice a long, almond-shaped building made of broken glass and broken ceramics. It's another building from our friend back in Kawakawa. This building, we find out, is called The Seed. It's shaped like a seed and had plants growing out of the roof. I think it's a great idea using up all the broken glass and ceramics and make a building with it.

As it is nearly our last day we decide to go to a Bird Rescue Centre. It's a great opportunity and we probably won't come across anything else as interesting before we fly home.

We find the place and it's quite big. There is even a miniature railway for kids. There is an office, but it looks closed. The building across from us looks open. We go in and there is a huge albatross…stuffed. A woman comes out from the other building and she says that all the stuffed birds have died naturally and are now on show to educate kids. The only Albatross I know was by Fleetwood Mac. She goes on to say that the kiwi is sleeping as it is nocturnal and the top guy is too busy anyway, so we won't be able to see it. A kiwi!

When are outside wondering what to do, a guy comes out and gets talking to us and because of Margaret's knowledge of birds, they get on great. We give him a donation for the centre. He says, "I can't have you coming all the way from Scotland and not seeing a kiwi." We are gobsmacked. We are going to see a kiwi after all.

The kiwi is called Sparky and he is a rescue bird, but he'll never leave because he only has one leg and couldn't survive in the wild. He lets us feed him and stroke him. We're told if we stroke him about the eye he'll nod over and sleep. A bit like Margaret. Not only are we seeing a kiwi, but we are actually feeding him and stroking him.

Sparky by the way is famous. Oh yes, he travels round the schools of New Zealand educating children. His handler was on the plane once sitting next to a German woman who was visiting New Zealand. She was unhappy that she was going home without seeing a kiwi. The lady got up to go to the toilet. While she was away, the handler opened the top of the box and sat it down on the woman's chair. When she came back, her face must have been a picture. On her seat was a kiwi peeping out the box. It reminded us of us. As it was the same for us in Australia: at the very last leg after nearly two months we see a duck billed platypus and an Aboriginal Sacred Dance and the haunting sound of the Didgeridoo. Now we get to see a kiwi. Once again we have managed to keep the best till last.

MAGGIE:

At the Bird Rescue Centre, we saw a giant petrol with a huge beak. It looked a bit like a dodo. Saw a Tui for the first time and a kiwi enclosure but no kiwi in sight. We spoke to a nice lady who said we had no chance of seeing the kiwi since it is too late as he is nocturnal, he is sleeping and everyone is too busy.

Five minutes later, a guy stopped to talk to us. We didn't realise it was the owner of the Centre. We spoke about birds in UK, Australia and New Zealand and I think he could tell we were genuinely interested and knowledgeable. He told us he would bring out Sparky the only kiwi in New Zealand who is allowed to be handled. We are very, very honoured and I feel quite teary.

Sparky only has one leg due to a trap catching him when he was young and he now is fifteen years old. He is taken all over New Zealand to do educational visits and talks. He is a national treasure. We were allowed to feed Sparky worms. Can't believe how lucky we were to meet such a nice kiwi and man!

PETE:

Well, seeing a real kiwi was a real bonus. We've seen a lot of things now and as we drive towards Auckland we discuss things we've seen. It's amazing. Where do you start? Just at that we see a sign One Tree Point. As it is getting late we decide to see if there is a camping area for the night. We enter the village. It is a well off looking area. We notice the street names are all American in some form: Florida; Bahamas; Hawaii.

We head down to the beach area and it's actually a lake. Very posh with lots of boats. Are we going to get moved on we wonder? We get a bit lost and we don't like the idea of people thinking we are cruising the streets. As we drive along the lake we come to a place called LA point. It's got a long gate which is closed. We're in an enclosed village. Like Orange County or something. Just up from it is some sort of car park, so we head in for a break only to discover that it is some sort of free camping area with a toilet. The top end of the car park has closed gates. There is only one way in and one way out.

When we wake up, we see that it is definitely a posh area. The garbage men came-cleaning everything and it wouldn't surprise me if they did this every day. Nobody gives us a second glance.

According to the trusty map there is a campsite off the beaten track called Baldrock View Farm; campers welcome. As we drive along the road we definitely are getting more off the beaten track, but we stick to our trusty map (it's been good to us so far). After quite a lonely drive, Margaret sees a sign Waipu Caves Glow-Worms. Well! you don't have to tell us twice.

We drive on a farm track for a while. It is raining. There it is: a big car park and there is only one other car there. There is an arrow sign pointing to the woods along the path, but that's all we see. We put on our waterproofs and head towards the wood. We're pretty quick in discovering that it is another rainforest with giant ferns, tall trees, streams, loads of insects, slippery, muddy tracks. Then you come across a cave, or should I say caves.

There is a couple of ways to get to them. I walk across a log. This is great. I feel like an explorer or a Native American or a Māori even. It's a small entrance so I had to watch my height and then it turns into a giant cavern. Then we start to see the glow-worms. The further in the more there are. I hear a noise and in front of us a big black shape. Please don't let it be an elk. Then a guy stands up holding his camera and tripod. A professional photographer I think. It's then I realise that glow-worms don't have ears. They won't run away. You can talk and they will still be there. Makes me feel better and I go in a bit further with the wee torches we still have from our encounter with the turtles on Heron Island.

Then we see them in all their glory. It's a bit like stargazing. There's more and more of them every time you look up. It's hypnotising. Then it happens-I slip into the murky water. I instantly think of bat shit, then I think of rabies and slide all over the place trying to get up. When I put my hand on something to pull myself up, it's all very slimy.

I forget for a moment where I am and I think snakes or even Anaconda. You don't get them in New Zealand, but then they said that about dangerous spiders.

I finally manage to stand up by holding onto something. That's when I realise that it's a stalagmite or stalactite or something like that. When I manage to shine my torch, I see loads of them looking like that scene in the film Alien where all the egg things were dribbling goo. The photographer is away and so is Margaret. I can see her torch heading out the cave. There is something that I can't resist doing. Very loudly I shout EHHHH OHHHH EH EH EH E EE OHHHH DAYLIGHT COME AND I WANT TO GO HOME.

MAGGIE:

We end up at One Tree Point free camping area. It's a very posh area and I feel very lucky to be able to camp here. We could stay as long as we wanted this morning since it is quiet season now. We left around 11.30 am. A Mobile Library was there. Lots of people were using it. Good vibe here. Folk are waving to us.

We headed up the coast to Waipu which I vaguely recognised. Then I saw a sign for Glow-worm caves. That's where I know the name from. The caves are free and the camping is free. We decide to go for it, even although it is raining. In fact probably the best time to go.

When we arrive at the car park/free camping area, it has what looks like a fairy glen. I like it. The caves have an eerie feeling about them.

I felt a bit like Arne Saknussemm from *Journey to the Centre of the Earth*. There were loads of glow-worms and it looked like the Milky Way at one point. Half an hour was enough for me. One hundred metres long was enough for me. There were slimy stalactites and stalagmites everywhere. Creepy.

Went back and had macaroni cheese and beans, then went to explore the fairy glen. We left some mud from Mount Cook and some shells. Then Pete decided to burn the black jacket he had brought with him. Glad to see the back of it as I never really liked it. Took loads of photos and there are some real corkers.

We end up at a farm campsite about ninety kilometres from Auckland. We are nearly at the end of our journey. Tomorrow we will camp at Red Beach campsite for the last time.

PETE:

We can't stay all day. Like it or not we have to go. When we go to the van, we notice that someone has had a fire, so now is the time for something else I want to do. Burn a jacket I got from someone. I feel as if it has brought me bad luck when I have worn it. I don't know why I even brought it with me. Here we are at the end of the other side of the world and I'm going to burn it. I make a small circle of stones and say something like you were too big anyway. I light the fire and keep turning it over and over until it is completely ash. That felt good.

It's time to go but before we do I notice some kawakawa and the leaves all have holes in them. We grab a small amount for our last fresh cup of tea to toast the glow-worms as we say cheerio.

Wolf Creek

PETE:

Well, we have only got two nights to go and so we've booked into a place near the airport. It's the Red beach campsite again. Once you put the code in and get past the gate it's got lots of streets and houses as well as static caravans and, of course, the red beach.

We have one more thing to do before we leave. We wake up and drive to Silverglade as we are going back to the Wolf Creek Brewery. I'm hoping my t shirt has arrived. It has and it's a cracker. It's black and on the back of the top it says in yellow writing Auckland, New Zealand and in the bottom right corner it says Deep Creek…on the front it has a silver coloured wolf with the shadow of a dream catcher blended in behind it. As I say it is a cracker. I also get a couple of cans of Deep Creek and it's still great. Apart from getting the t shirt and visiting the brewery there is not much to do here. We head back to the campsite and have an early night.

Well this is it. Our last day. We go for a walk on the red beach. I make my peace sign in the sand with a stick. It's getting quite dark now, so it's now our last night. There is just enough light left to ask Margaret to sit on a rock next to a smaller rock. I write on it and take our last photo. I show her the photo. It says:
Pete and Maggie
New Zealand
30-05-2019

THERE AND BACK AGAIN

MAGGIE

We are back up in the air heading home to Scotland. The past couple of days I could feel that I was ready to go home. Not homesick, but just ready. It will be lovely to see our families again. The weather took a wee turn for the worse too. We have had so much good luck on this trip. The weather is an example. Blues skies nearly every day. Very rarely did we suffer from sweltering heat in Australia. We lived like they did: got up early, in bed early. It was gorgeous. Our timing has been great just coming out of summer and into autumn in Australia and just coming into autumn in New Zealand.

There is definitely a paradox with New Zealand. The South Island possibly has the most beautiful and diverse scenery in the world, yet a lot of the young people in the South seem disgruntled, angry even. Their native wildlife, both fauna and flora, are suffering terribly and the government are literally poisoning paradise.

Mind you there are only four per cent of the Māori population living on the South Island. Maybe that's what makes the difference. The few we met were so pleasant. The North Island was much friendlier, probably because there are ninety six per cent of Maoris living there. I remember meeting a Māori guy in Australia and I asked if he missed home. Not a chance, he said, I much prefer Australia. This is the difference I notice about the people of Australia and New Zealand. No matter what life throws at the Aussies they seem to be able to keep positive, keep laughing and joking and just get on with it. I noticed that the young people also have great manners. There was much kindness in Australia. Having said all that we have had such a wonderful adventure. I feel so much gratitude. I'd love to come back again one day.

Just now though we have been there and now it's back again to Scotland. The word Brexit comes into my head. Brexit, Brexit, Brexit. Every day for nearly three years it was all the media could talk about. It slowly poisoned the goodwill of the people. For the past four months, we haven't watched TV or read papers. It has been bliss.

As we fly over this enormous, diverse, wonderful planet, I feel privileged to be alive and be part of it. As I watch Mary Poppins Returns, I think what we all need is some of Mary Poppin's fantastic mixture of common sense and magic. Maybe she is teaching us just what we are capable of. If we just believe.

PETE:

Although we left New Zealand on a high, we were ready for back home (just). Knowing that the flight was twenty one hours or more, I deliberately hung off from sleeping. Glad I did because I fell asleep quite early on, for a while. When I woke up a few hours later, I looked out the window and saw a mountain range. *That looks like India*, I thought. In fact, we were still flying over Australia. That's how big it is. It takes five hours and fifteen minutes to fly from Brisbane on the East coast to Perth on the West coast. With that I nodded over again.

This time when I woke up and saw another mass of land, I checked the radar screen in front of me. It said Colombo not Colombia. I asked the Stewardess and she confirmed that we were indeed flying over Colombo in Sri Lanka. MAN, I've always wanted to go there. All those weird animals…or is that Madagascar?

Anyway, we get home in one piece. We pick up our campervan Gabie.

He will be our home in Ayrshire for the next few weeks while we work, rest and play. We are getting ready for the next part of our terrific journey: Orkney and Shetland.

Our van seems huge compared to the small ones we have had in Australia and New Zealand. We have got ourselves Icelandic Wool blankets and other woolly things to help us battle the elements up north.

WOW! That time already? It's Sunday 1 September 2019. It's time to go to Orkney and Shetland…but first…we need to get there.

MAGGIE:

When we landed back in Ayrshire, I must say I found it very difficult to acclimatise. It was June, so it wasn't the weather. We had been away for four months. I came back with fresh eyes and fresh ears.

The litter in Ayrshire already annoyed me. When I got back, after being in such beautifully clean landscapes and beaches, it seemed even worse than I remembered. I am at a loss to know why the people here have such little respect for their environment, nature, their fellow human beings and ultimately themselves. I heard a famous rock star, who happens to be married to a Scot, accuse Scotland of being a third world country. This is a guy who was brought up on a rough council estate, so it must be bad. I quite often find him unnecessarily cruel with his words, so I was alarmed when I found myself agreeing with him.

216

In Ayrshire, the fallout from long term drug abuse is evident everywhere. People look ill and the countryside looks ill. There is an air of neglect all around. Children are crying out for parents to have some time for them and some boundaries. Delinquency abounds. Also wherever I go people are obsessed with politics. I used to be like that myself. I am so glad I have managed to come off that particular addiction.

I try my best to keep my vibes up. Living in the campervan really helps. The peace and quiet, being near to nature, no TV etc. It is so easy to be dragged down. Let's face it, a lot of people like to drag you down to their level. Subconsciously or consciously, it makes them feel better about themselves. Can't wait to head to Shetland.

Shetland is a place I have wanted to go to for a long, long time. I even have a feeling that my ancestors might have come from there. At last, our leaving date has arrived and we can head off into the unknown once again.

Care No More

PETE:

A good start to our journey as leaving at 8.30pm meant it was a nice quiet drive up through Glasgow towards Stirling. We carried on towards Perth. We eventually stopped at a lay-by for the night.

Refreshed, we decide to visit Scottish Fruit Winery—Cairn O'Mohr. After a few glasses, it is pronounced Care No More. We speak to the owner who has been to Australia and we have a good chat. It turns out he goes to the Shetland Isles. He is a bit of a twitcher—a bird spotter. They're the folk that get up at the crack of dawn to see a rare lesser spotted African goose that's been blown off course and somehow lands on the Shetland Isles.

After a few travelling tips, we say goodbye. We head into the cafe next door and get some of the three soups. This should be world famous. It's a great idea. Instead of wondering what type of soup to choose, you get three small bowls of them all.

I even end up getting a tee shirt in the shop and wear it with pride. It's time to go and give a toot toot, honk honk.

MAGGIE:

After saying our farewells, we set off just after 8.30 pm. The drive was so smooth it seemed no time until we were up past Stirling. We saw a lorry in a lay-by and thought—that will do for us. Pete had bought a bottle of Prosecco to celebrate. He chose to play the song Me and You and a dog named Boo by Lobo. I chose On the Road Again by Willie Nelson.

We had a great sleep even although the road seemed extra-ordinarily busy. We decided to head for Braemar. I remembered the Cairn O'Mohr Winery was somewhere nearby. Lo and behold, as luck would have it, there it was on the Perth to Dundee road on our way to Braemar.

We were delighted that the owner was able to come out for a wee chat. It being Autumn he was very, very busy. He is a real character. His Winery is well worth a visit. There are loads of big wooden Easter Island/Māori like carved faces dotted around under the title 'Heids in Weeds'. He is a keen bird watcher and has a list of sightings. The shop has loads of delicious wines and we got to taste a few, before treating ourselves to a bottle of Back O The Door. This is made up of leftover wine from each of the different batches.

The cafe next door had the best coffee I had tasted in ages, in fact since Brisbane, and the trio meze of soups is a novel idea. Just wish I could have done the same with the cakes as there are so many amazing cakes to choose from.

PETE:

We are heading up the road to a place called Braemar. I keep thinking the name's familiar. Then I realise it is where the Highland Games are held. In fact, it's The Royal Highland Games, as the Queen and other Royalty attends every year. In fact, they have been attending for well over one hundred and fifty years since the days of Queen Victoria. Also celebrities such as Billy Connolly and USA Presidents attend. Not sure if Donald Trump is attending this year.

Braemar is a lovely, litter free village, nestled in rolling hills of purple heather. Beautiful. There is also an area for free camping. The Highland Games are actually on next week and the village is buzzing preparing for Her Highness and Co. We won't see them as we've got to reach John O'Groats soon, to catch the ferry to Orkney. Tomorrow we're going to Aviemore. Home to the Cairngorms—dangerous and jaw droppingly beautiful. A skier's paradise. Tonight we open a bottle of Cairn O'Mohr wine which is delicious and gives us a good night's sleep.

MAGGIE:

Arrived at Braemar around 3pm ish. We got to see some blooming heather on the hills although it is just past it's best. Braemar is a lively wee traditional looking town with a calm vibe. It restored my faith in Scotland. There was no litter.

We were made to feel very welcome. We bought some lovely sausages from the local butcher. Also some Stornoway black pudding, and some pork and apple burgers. He told us we were welcome to camp in the car park overnight as there is a toilet. He reminded us it would be the Highland Games gathering the following Saturday when fifteen thousand people descend, including the Queen.

There is a huge Hotel which tourists will either love or hate. It has an enormous selection of animal heads on the wall including wildebeest. There are lots of interesting nooks and crannies. The Stag Bar had an actual stuffed deer hanging from the ceiling with wings, as if flying. Eeek!

We had a cosy night with our fruity Back O The Door wine. It made me think of the word Hyggelig which the Danish use to describe the cosiness that winter can bring, especially if you embrace all that quality time with the people you love and of course yourself. The wine is delicious. Going to keep the other bottle for Orkney. Hic.

Woke up to nice weather and we headed off via Crathie to have a peek at the church and at Balmoral if possible. Turns out it's off the road and we head towards Ballater and then over the hills via Tomintoul to Aviemore. The sun came out for us and it was a beautiful run with the heather still blooming in some places. We had our Braemar fry up in Tomintoul.

We visited my old skiing stomping ground-Grantown on Spey. Lovely thriving town and looks grand with four well preserved Hotels. We visited the Grant Hotel for a look. It is not the musty, dimly lit place I remember from 1980s and only has three stags heads on the wall. I have fond memories of those skiing trips.

Heading through my beloved Cairngorms to Aviemore, I never, ever tire of it, even in the rain. We are lucky as it stays dry and we head to the Winking Owl pub once we arrive, purely as a nostalgic trip of course. We were so cosy in bed, we slept until noon.

Once we had surfaced, we went for a walk around town and the Heavens opened. Pete got drookit as he hadn't brought a jacket with him. We dive into

The Old Bridge Inn which is a traditional pub. We liked it very much and had a Drambuie to help us face the elements. It certainly gave us a glow.

After drying off, Pete took me to a Charity Shop. He found a camping light which runs on batteries. We notice it is made by a couple in Wales who loved camping so much they started a successful business. That's the way for us all to live our lives if we can: Do what you love.

I don't know how this came to be, but there is a wee sticker on my knickers. It said, 'Made from the finest ingredients'. That gave us a right laugh.

PETE:

After a lovely drive up through Tomintoul, it's dark when we reach Aviemore. We head over to the Winking Owl pub getting warmed up and more importantly we get the laptop out and set to work.

We get our first page written about our Australian and New Zealand journey. We feel real good that the book has now started. While we write Part One, we are also writing our diaries for our present journey:

Part Two

Orkney and Shetland

Last Place on Mainland, Scotland
Groatie Buckies

On our way to John O'Groats, we decide to have a fish supper. We had stopped in a wee village with a picturesque harbour. We saw a sign in a window which said BEST FISH SUPPER IN SCOTLAND. The very place for us. As I walk in there are two men sitting at a table talking. "No Wattie, you must go straight hame, nae hinging aboot. In fact, phone me so as I ken yir fine. Right then Aff yi go." "Hello sir whit can I get yi. We do a braw fish supper."

As he/she stands up it's very hard to tell. She has rosy cheeks like a doll and an apron like a granny. He/she checks to see if the pal is away. "Ken this, the bloody polis!—we ended up at the polis station. A cannae get ower it. Polis, ae aw folk! Here's yir fish suppers. I'm glad you didnae hiv fish cakes, a widnae hiv gave yi them. There no that gid. These fish suppers are awfy gid yill enjoy them. Here's salt and vinegar. Help yersell fur nuthin, but no the tartar sauce, you'll need to pay for it if you want it. A cannae get ower it. Bloody polis. A better phone him. Bye now enjoy your fish, the best in Scotland."

Well they are certainly good and plenty of them. After our best fish suppers, we head for Wick and it's not far to John O'Groats.

MAGGIE:
We are on our way to Dornach via Carrbridge and Dulnain Bridge. Carrbridge had loads of lovely wood carvings and the oldest bridge in the Highlands, built in 1717. It also boasts really nice toilets. Simple and effective, but so important. It is way more hospitable to campers up this way. In Ayrshire, probably due to the amount of drug abuse and vandalism, you are lucky to find a public toilet open.

After a visit to Dulnain Bridge, we head up the road towards Dornach.

I have only been to Dornach once way back in 1981 and I had forgotten just how lovely it is. There is a feeling of bonhomie, no litter, lovely beach. A town that believes in itself.

After wandering around for a few hours, it's time for the road again. We kept seeing a rainbow all the way up the coast. Not sure if it is the same one, but it was really bright and would appear and disappear and re-appear as if someone was switching it on and off. It was magical.

We stopped at the Hill of Many Stanes. There are many theories but, no-one really knows why they are there. My feeling is that it is where many folk were killed in a battle. Pete felt it was something to do with the stars. I prefer his theory.

Kept on trucking towards Wick passing through lots of lovely wee villages and towns. We spotted a we village I recognised from a TV programme. It has a beautiful wee working harbour. Thinking about the rainbow, as we stroll, I ended up singing one of my favourite songs as a teenybopper in the early 1970s—David Cassidy's Daydreamer…

After my walk down memory lane, we spotted a Fish and Chip shop. The wee person behind the counter was a real gem of character. Pete did a sweet impersonation. We both ended up talking like Mrs Doubtfire for the rest of the day and night.

PETE:
We finally reach the famous sign at John O'Groats. We stand and get our photo taken. There are arrows pointing to different places and to my amazement there is an arrow pointing straight down saying New Zealand. It brings back real good memories.

Rather than just having a photo taken and leaving, we decide to stay the night. After all, that is what campervans are for. Next day we walk along the braw beach and eventually out of sight of all the tourists. As we are walking we come across a few locals searching for something. I ask if they have lost something and they tell us that they are looking for very, very small shells. They are called Groatie Buckies. Groatie as in John O'Groats and Buckies as in, well Buckie Shells. Whatever they are. A woman gives me one for luck and then I eventually find some. You are supposed to thread them together and make jewellery from them. They are so tiny that it must be hard to do.

We do our last piece of shopping in a small shop by the pier before we head to Orkney. As we look around we notice there are three bottles of Cairn O'Mohr mulled wine on a shelf. They are a good price, so we take all three. Cheers that should keep us warm.

MAGGIE:

When we arrived at Wick, it looked a bit grey, drab and down on its luck. The shops could be doing with a lick of paint. Kept on driving and finally arrive at John O'Groats. Despite a very strong wind, we go out to the famous John O'Groats sign for the obligatory photo. Feels as if we are in danger of being blown to the North Pole. The car park allows overnight parking and charges £2 for twenty four hours and so we stay for the night. Love the Highland hospitality.

In the morning, the weather, as predicted, had calmed down. We can walk upright without being blown away. We noticed that cars would pull up, everyone would jump out, go to the signpost, take some photos and be off and away within five or ten minutes. I reckon most of them are doing the North Coast 500 which has become a bit of a ticky boxy exercise for some folk. There is much more to John O'Groats than first meets the eye, if you take your time to look. It has a good vibe. We decide we are going to savour the visit and stay for another day.

The beach is beautiful and we decided to walk a few miles to the east to see the rock stacks at Duncansby Head. As we walked along the coastal path, we came across a few folk looking for something in the colourful shingle. Intrigued we asked what they were looking for. The tide had not long gone out and they are searching for some sort of wee sea snail creature called a 'Groatie Buckie'. It's actually called a cowrie shell. In Caithness, they are considered to bring good luck to those who find them. A lovely lady gave Pete one for luck. They are tiny and beautiful. No wonder people make jewellery from them.

The lady explained that all the people we see are local and it's a sort of family pastime. All ages children to Grannies like to collect them and make things from them. She also told us that this is the best beach in Scotland to find a Groatie Buckie. It is lovely to see families out together enjoying nature. I had no luck in finding any, although we spent ten or fifteen minutes looking. I had a wonderful time. We are living the dream, making time to do this and in such a pretty location. So glad we were not in such a hurry to tick a box.

We left them on their mission and we continued on our walk and came across a beautiful cove with a different type of beach. There was no shingle, but there were lots of bigger stones being taken in and out the waves like flotsam and jetsam. The rhythmic, rattling sound was sort of hypnotic and captivating. With its magnificent power and natural sense of infinity, the sea is a place I find to be timeless and healing. I can sit for hours watching and listening to the waves. I suppose it is a form of meditation.

We made it to the rock stacks: there was a small, medium and large. Definitely worth seeing. There is a nearby lighthouse and plenty of birds. The stacks are a bit like the pictures of the Old Man of Hoy. Speaking of which Orkney is beckoning. So glad we took the time to explore here first. On the way back, we tried to find some Groatie Buckies. Pete decided just to focus on the tiny unique shell pattern and then he proceeded to find one and two and three and eventually I got the hang of it too.

PETE:
It's time to go to Scrabster as that is where we get the ferry to Orkney. After having breakfast in a car park come nature reserve area, we head on up. As we are waiting on the ferry we spot a huge bird. It's a raven which seems appropriate as when the ferry comes into sight we see a huge Viking painted on the side of it.

As we collect our tickets the booking office is empty. Not surprising as we are quite remote. It was either that or it was the humdinger I dropped.

When we get on the ferry, I get my second realisation just how remote we are. To get to Orkney we have to cross the North Atlantic. Maybe we'll see an Orca…

When we arrive at Stromness just up from the ferry port, the first thing I notice is a shop selling rolls and mince. No prizes for guessing where I'm heading. Our phones are as dead as a dodo, so we can't take pictures.

We've found a nice little spot at a beach. Margaret is desperate for a fire, but it's much too windy. We notice a poster advertising a talk on Norway, the Vikings and their connection with Shetland. Margaret is wanting to find out her heritage here. She is convinced she is a Viking, so we better go. Her maiden name you see is Dalziel. It's pronounced Dee Yell and one of the Shetland islands is called Yell. Also they pronounce 'the' as 'da' The Yell—Dalziel—Of the Yell—get it? Well, that's her theory anyway.

Off we go to the Royal Hotel to hear what they have to say. To our surprise for a small donation and some raffle tickets we get to hear an interesting talk on folklore and even better, sample Orkney hospitality. At half time, we got some lovely smoked salmon oatcakes, Orkney cheese and all sorts of interesting traditional cakes and goodies. We didn't win any raffles. The chairperson's son, who owned the pub, won most of the prizes but his dad put the tickets back in to try again.

MAGGIE:

We head off towards Thurso, which is a bit big for our tastes. Eventually we find a lovely camping area near a wind farm. Only found out how close when we got up in the morning and saw them towering over us, like gentle giants.

It's a beautiful morning. We give the blankets a shake and head off for the ferry. There is a huge black raven circling above us. There's only one, so it can't be Huginn and Muninn the two ravens from Norse mythology, the ones that report back to Odin. It does feel as if we are heading into Viking territory though.

The ferry journey takes ninety minutes. We were disappointed that the ferry didn't have a commentary as we nearly missed The Old Man of Hoy and just caught a glimpse of the back end of him. We then sailed through the famous Scapa Flow and we see could Stromness ahead. When we drove off the ferry and took a wrong turning, I got to see how quaint, charming and narrow the streets are. Luckily it was a Sunday.

After finding a car park, we strolled through the streets and then headed off to a lovely wild beach for a walk. Going back to Stromness to listen to a talk at the Royal Hotel on the folklore of Norway and Orkney.

We arrive to a warm welcome. We buy some raffle tickets and make a donation at the door. The talk is interesting. A local storyteller, a tradition much valued up this way, showed us slides of the northernmost point of Norway islets and how story/myths in Orkney are very similar to theirs—about selkies and trolls and involve the sea. Further south in Norway the tales are more Hans Christian Anderson and Brothers Grimm.

We have noticed that there are a lot of Norwegian flags flying on Orkney. They feel a real affinity. I have also noticed that the Shetland flag is like the Scottish Saltire with a Scandinavian twist. The Viking influence seems to be everywhere.

We sat next to a lovely lady who had lived in the same house for sixty three years. She is well into her eighties now and tells us how lucky we are to be able to travel together as her husband died when he was only fifty five. Mind you we were delighted to hear she was shortly going on a cruise to Switzerland via the Danube. The Orkney hospitality was amazing and again the Scandinavian influence was evident. We got a smorgasbord supper with herring, smoked salmon, local cheese and oatcakes, followed by cakes including a local almond and apple cake. Thank you kindly.

PETE:

We went to a beach campsite run by the council. For a shower, we need 20p pieces and we don't have any. I went to a local shop and this is the conversation between me and the girl in the shop|:

ME: Hi, can you give me five 20p pieces please for a £1.
GIRL: Sure, no problem.
Me: Thanks.
GIRL: So? You don't want 10ps or 5ps or any other coins, just 20ps?
ME: Yeah.
GIRL What on earth do you want all those 20p pieces for?
ME: It's for a machine which only takes 20p pieces.
GIRL: What? In here?
ME: No. It's at a campsite.
GIRL: For a campsite? You need 20p pieces for a campsite?
Me: No, it's for a shower. They take 20p pieces and I don't have any.
GIRL: Yes you do. I've just gave you five.

They're a funny lot up here.

We get settled for the night, have our 20p shower and get ready for tomorrow. We are heading to the capital of Orkney: Kirkwall.

When we arrive at Kirkwall, we go to the Magnus Cathedral. It is huge and very impressive. Built around nine hundred years ago it is in great nick. Well worth a visit and it's free to go in. They hold music concerts in it and the acoustics are bound to be brilliant.

Here is a bit of interesting information. The Cathedral belongs to the people of Orkney and is known as the 'Light of the North'. It was founded by the Viking Earl of Rognvald in honour of his uncle St Magnus who was martyred in Orkney. Someone told me the caretaker's name was Ragnar of the Hairybreeks. It would be great if it was.

We moved on to see the famous Orkney Standing Stones. Of course we got lots of pictures standing next to them. They are awesome and it's mind boggling how they got there and it's even more mind boggling why? I can easily picture sun worshipping going on here and huge bonfires. It looks great at sunset. Very mysterious...

Then it's off to another great historical site: Skara Brae, the neolithic Pictish village. When I see it, I am more than convinced that Tolkien must have used it as his Hobbit inspiration. These are real life wee hobbity houses in the hillside and everything is for small folk. The Picts were seemingly small, friendly, peace loving people and stayed here until they were invaded by the Vikings. When you look at the size of Shetland ponies, the Sheltie dogs, even the sheep etc., they are all small and yep I think it inspired him. Even the Elven language in the books is based on Viking rune stones. Just to convince me even more, I asked the guide if Tolkien came here to visit and he said yes, but even better two weeks ago Gandalf was here. Yes, Sir Ian McKellen had been visiting two weeks ago. That's it. As far as I am concerned Tolkien definitely used this place to base his Hobbiton.

MAGGIE:

Had an amazing day at the Stones of Brodgar. We managed the guided walk which was very interesting, but no-one really knows what went on all those centuries ago. The guide did tell us that big things, in the archaeological sense, are happening. It is now realised the people from those times were more advanced than previously thought. There is evidence that people travelled from all over Europe and Britain to take part in whatever it was they got up to. Maybe it was like the Glastonbury Festival for its time. Now that would be cool.

We went on to Skara Brae. The place looked like Hobbiton. When asked if Tolkien had visited, the guide confirmed that indeed he had. We were told that Sir Ian McKellen had visited here just two weeks previously which was a bit deja vu like as Harrison Ford visited Hobbiton, New Zealand two weeks before we visited there.

It feels as if we are following in Tolkien's footsteps without trying. For instance, when we went to Swiss mountains two and a half years ago to visit Alfred Vogel's Swiss Herbal Museum, we also visited the astounding Lauterbrunnen area (the name means many springs). This area is truly magical.

We had come to see the waterfalls and there are seventy two waterfalls to be exact.

Our favourite was the Trummelbach falls. This is ten glacier fed waterfalls inside a mountain which were made accessible to humans in 1913. Breath-taking. It turned out we saw a poster advising that Tolkien had visited them around a century earlier. I reckon he took inspiration for the Hobbit mountain from there. Now here we are in Orkney and he has been here too, albeit many years before. It was good to go 'back in time'.

PETE:

We drove up to a spot overlooking the sea which has a car park. When we got out to investigate, we find out there is an old Pictish settlement across the sea on a small piece of land. We decide to go to visit it and luckily we read the notice that tells us to PLEASE BEWARE OF TIDES. The tide seems to be quite a way out and we manage to walk over the crossing without falling in. A guy passing us tells us we have about an hour. It is well worth it to see the well preserved small settlement.

A lesson in the way of life of the Viking settlers (the Picts were long gone). It is so well structured I wonder where they got their technology from. I see another sign a wee bit away from me. I decide once again to investigate. I get a bit of a jolt when the sign tells me if you get stranded please don't panic, either call the coastguard on this number or stay where you are until the tide comes back out again.

When we look out to sea, we decide to get back pronto as an hour will be too long. We make our way back and it's a good job. We are walking over trying to keep our balance and the tide came in so quick the water is splashing over our boots. We walk a bit quicker but have to wait till the wave recedes before moving on again. From the car park looking down, it might look like we are walking on water.

When we finally climb back up to the car park and look down across the water to the settlement, I try to imagine the little place in full swing back then in Viking or Pictish times. I couldn't help thinking that they'd picked a really secure and safe place to build a small village away from danger. I wonder what happened to them?

MAGGIE:

It was also recommended that we visit Birsay to go across the causeway (it's a bit like Lindisfarne and depends on the tide). We had to be quick because the tide was coming in and we were warned we only had an hour or we would be stranded.

Good job we went back in half an hour because the water was starting to come over the walkway and we just managed back in time. We then went a walk in the village and visited St Magnus kirk. There were lots of information about how Magnus gained Saint status. He had to wait until twenty one years after his death to be given this status even though there were lots of stories that he performed healing miracles. Seemingly a clergyman who had refused to give the saintly status previously, lost his sight and he got it back after praying to St Magnus. That convinced him.

What a day we have had. We had nipped into a Brewery on the way back down the Island. Enjoyed a conversation with a very pleasant young Australian lad. He and his partner have a two year work permit here. We got to reminisce about Australia. We told him about the town Glen Innes basing the stones there on the ones at Brodgar.

PETE

Back at Kirkwall we see a poster for Kraken dark spiced rum outside a pub. We decide to give it a try. The pub is called Torvhaug (an old Norse word for peatstack). The word Kraken immediately conjures up one of those Jason and the Argonauts films where Jason tries to release the beautiful princess and the baddie King shouts out RELEASE THE KRAKEN—a big monster which looks like a cross between Godzilla and King Kong. It swims underwater and snarls and roars at the princess till Jason holds up the head of Medusa and the monster turns to stone. A bit like us if we don't get out of the cold.

Sadly there is no sign that it is named after a mythical Greek creature. It does have three televisions, a jukebox and a coffee machine behind the bar. I order two Kraken with ginger and lime.

The guy behind the bar is friendly and gives us information about the Churchill Barriers. He explains they are huge lumps of concrete lined up underneath and above the water along the entrances into the gateways along the shorelines of Orkney. Churchill ordered these concrete monoliths to be strategically placed along these coastlines and our barman storyteller goes on to inform us that he also deliberately sank boats and left them. Why? Well quite ingenious really, it was to stop invading submarines from silently sneaking in and taking over the top of Scotland. To be honest I didn't realise how much Orkney was involved in the War. My dad was in it, so it's not that long ago. Top marks to the Prime Minister for being clever and foreseeing Hitler's strategy. We thank our reliable barman. I have a feeling we will be back at the Torvhaug.

MAGGIE:

It was another wild night with lots of rain. We were cosy with our blankets. The next day it felt exhilarating walking up to more ancient sites. The weather might have contributed by blowing cobwebs away.

When we come back to Kirkwall, we notice a poster outside a pub for Kraken dark spiced rum with ginger and lime. The very thing on a windy day especially the ginger. The guy behind the counter was very nice. He gave us a lot of good information about Scapa Flow. I didn't realise the extent of the German threat to Orkney during the Second World War. Part of the Churchill barriers at Scapa Flow are scuppered boats which were put there to stop submarines entering. Eight hundred men had tragically lost their lives to a German submarine. The boats are still there to this day. We decide to go and see for ourselves tomorrow.

I am glad we visited these different sites and took the time to walk up to the Stones. The island's history is fascinating and there are a lot of mysteries which no historian can quite explain. There are literally tombs all over Orkney. It gave us a feel for the scale of these mysterious gatherings and I can't help thinking about the cosmos.

Time to get cosy and we park at a RSPB recommended site. I asked Pete if he liked Kraken and he shouted ODIN or more like OOOOOOOOODIIIIIIIIN. He actually sounded more like Tarzan.

The Marie Celeste

PETE

We got to see the Churchill Barriers without trying. At first, I thought it was a very long bridge in the distance. As we get closer I see that in fact the bridge is actually a road made through a line of huge boulders. I notice some old sunken wrecks, the front of the boats sticking up like periscopes. The barman wasn't exaggerating there are loads of them. We sit and look at them for a wee while. I wonder what it must have been like living here then. Wondering if a submarine would suddenly appear or a burst of gunfire or something.

Orkney is a really fascinating place full of history. One such place is called the Italian Chapel. It's made from two old Nissan war huts and homesick Italian POWs painstakingly painted the walls as if they are made of bricks. There are angels, saints, cherubs. It's hard to believe they are not real and that the building is made of plain concrete. You really do think you are in a real chapel it is such a high standard of painting. In fact, if they were still here, they could do our bedroom.

When we came out of the amazing Italian Chapel, there is a small distillery at the foot of the path, so we had to go in and investigate. To our surprise it is a small rum distillery with a few different flavours which we get to sample and which are delicious. They also have fruit wine. We get an Orkney white wine with hints of gooseberry as we are getting ready to catch the Ferry to Shetland on Saturday.

Driving back up the road we come across what looks like a scrapyard. It's full of lorries, tractors, cars and all sorts of things. I ask Margaret to stop as I haven't forgot that we need a wheel trim and one that fits. There is a couple of houses next to it which I assume belongs to the scrapyard.

I knock on the door but no answer, so I knock again and still no answer. There is a house nearby so I go over and knock on that door. A lady appears and I explain about the wheel trim having fallen off our van. I ask if she is the owner and would it be ok to look for one?.

"Oh no I'm not the owner. There has been no one there for well over two years."

"But, I say, there is a couple of cars in the driveway and it looks as if someone lives there."

"I can assure you," she says, "there is nobody there. Away in and see what you can find. Nobody is bothered."

I thank her and then think nice one as I wander round the back. There are sports cars, old ford escorts, a jaguar, Morris minors and all sorts lying derelict and rusty. It's a car mechanic's dream. I would love to repair and get one of these things going again. However, I'm on the look for a wheel trim. Well! there's plenty, but not one our size. Our van is a converted taxi and it's got rather big wheels. As I make my way back round the place, I can't help but notice the back of the house looks as if someone has just gone indoors. Everything is lying around as if they have been gardening, a dog bowl at the back door, a small window open. It's hard to believe that there has been no one here for two years. When I look in the window, I see all the furniture and even plates on the table. It wouldn't surprise me if I saw Goldilocks sitting at the table and the three bears upstairs sleeping.

I could spent all day looking around a scrapyard, but I have been here long enough and Margaret must be fed up waiting. I grab a wheel trim in the hope that it fits. It doesn't, so I put it back. It's time to move on. As we drive slowly away we pass the second part of the house and there, in the window, in huge letters staring out at us, are the words BE KIND.

Who would leave everything, including a very well stocked scrapyard, a house full of furniture, a dog bowl? Have the words BE KIND in the window and just disappear? It was like something out of the Marie Celeste ghost ship. All we can do is drive on and I think, well, that's Orkney for you.

MAGGIE:

Next day we visit the Churchill barriers and the Italian Chapel. It was the Italian POWs who built the barriers and the Chapel. We can see lots of scuppered boats sticking out of the water. The Italian Chapel was breath-taking. The paintings and the sculpture of Jesus were so lifelike. When we were there, a group of Italian tourists came in which was nice.

After driving around sightseeing and stopping at a lovely wee place called St Margaret's Hope, we drive back up towards Kirkwall. We had lost a wheel trim from Gabie and decided to look out for a scrap yard or garage which might have a spare one. We came across what looked like a scrap yard and stopped to ask. No one answered the door. It turns out that the house was abandoned a few years ago and no one lives there although all the furniture seems to be sitting in the living room. Why would a family just up sticks and leave absolutely everything at such short notice? There a huge sign in the window which says BE KIND. Another Orkney mystery.

We have been lucky with the weather today, but torrential rain comes on just as we are walking along the harbour back in Kirkwall. In fact, when we run towards the campervan, we turn a corner and the wind is so strong it blows the rain and in a split second it feels like someone threw a couple of pails of water over us. We are drookit.

After a change of clothes, we head back to the pub for a cherry rum coffee which our friendly publican had recommended the other day. The guy tells us that the weather forecast is terrible and the ferry tomorrow has been cancelled as well as the Scrabster/Stromness ferry. Seemingly this happens a lot at this time of year. We decide to head up to another pub for some live music jamming session.

When we get to the pub, we have a couple of southern comforts each with ginger ale which takes me back to my clubbing days in the eighties and my favourite tipple a 'Gatsby': southern comfort, ginger ale and lime.

The musicians show great musicianship picking up riffs as they go along. A wee guy came in with a mandolin and proceeded to do a couple of Bluegrass songs. One called Rain Please Go Away was my favourite. Each musician takes a turn to choose a song. Great blues as well as some traditional reels. We left about midnight and they were still playing just for the love of it. Noticed that there were lots of lone travellers and that they sat with one cup of tea all night. It made me think how lucky I am to have Pete by my side, supporting each other, sharing our adventures and making each other laugh.

PETE:
We have noticed lots of small stone wall circles all over the Island and find out they are called planticrubs. They are high enough for crops of vegetables or plants to be sheltered from the fierce winds up here. It's so windy it's no wonder the locals are well wrapped up and check the weather constantly. When I look in the mirror, I realise why they all wear hats. I look like a cross between Albert Einstein and Beetlejuice. It doesn't help when trying to run round a corner in torrential rain, we both freeze with shock and cold. It's as if someone in front of us has thrown ice cold water over us. We are soaked to the skin. I think of the Churchill Barriers. I don't think they could even have stopped it. It was a reminder that up here in Orkney and the Shetlands the weather can be real harsh.

After changing our clothes, it's time for a hot drink. I knew I would be back at the Torvhaug! We order a cherry liqueur coffee so that should do the trick. The guy behind the bar had recommended it as his favourite coffee. We watch him expertly rustle them up and it's even better than it sounded. Straight away I wish we had ordered a double. He goes on to tell us that the weather is going to take a turn for the worse. I had already thought it had, but seemingly this is normal and it is going to be real gale force winds tomorrow. A guy in the pub looks up his weather app and right enough the ferry has been cancelled tomorrow night.

Jings, crivens and help ma boab whit are we gonnae dae? Well, we ken whit we are gonnae dae. There's a pub up the road which we noticed had a sign in the window, 'Jamming Session—All Welcome—Bring your own Instrument'. I get out my air guitar and off we go.

We walk up to the bar and it's quite small with only a few beers on tap. No real ale. We both spy a bottle of Southern Comfort. Man! that takes me back to when my daughter was born forty three years ago. I remember all my mates buying me a round and I ended up losing my voice. Margaret says it used to be her favourite drink as well. So it's a unanimous decision and we order a couple. The girl asks what we want in it and we both say ginger ale at the same time.

As we sit down to wait on the jamming session to begin the musicians are still coming in. Most of them have fiddles, but also banjos, guitars, mandolins. They start with a couple of traditional tunes then some bluesy tunes. I am loving it and the old foot is doing that thing I do as if I am drumming. They play a song I know by Canned Heat so that impresses me. As they are playing a guy and a girl walk in and join them mid tune. Love it. As I look round it dawns on me that me and Margaret are the only two drinking. All the rest have cups of tea. They are tapping their feet, but somehow it looks a bit funny. It gets even funnier when the guy who runs the pub starts singing a song about Rye Whiskey:

Jack o' Diamonds, Jack o' Diamonds and I know you of old
You've robbed my poor pockets of silver and gold
It's a whiskey, you villain, you've been my downfall
You've kicked me, you've cuffed me, but I love you for all
It's a whiskey, rye whiskey, rye whiskey I cry
If I don't get rye whiskey, well, I think I will die

After the song, he then looks at the bar and says, "C'mon folks, get up to the bar and get yourself a drink!"

At least, we have both had a couple of Southern Comforts each so I hope the guy has noticed.

A wee guy gets out his Dobro, that's the instrument for bluegrass stuff. It sounds great. Even the tea drinkers have both feet tapping.

He plays a song called Rain Go Away, Don't Come Back till Another Day. It's just fab. I write down the name of it so I can look it up. During the interval I go to the toilet. It's up the stairs and as I go up I notice a lot of pictures on the walls of four girls with musical instruments. I find out that they are called Fara. Seemingly they are now famous in Canada, America and even the Far East. They are real life local heroes and every now and again come back to Orkney and encourage the local young ones.

Back at the bar Ol' Mr Rye Whiskey is doing another gravel voiced blues song with the guy next to him playing a fantastic tune on his guitar and fits in with it. It's great fun listening to them all joining in…real jamming…

As we are leaving the band is still playing and it must be about midnight. It was a great night. We head down to the van feeling good and glad that we went.

We wake up nice and fresh and I can't stop singing if I don't drink Rye Whiskey I surely will die. We head to the Tourist Information Centre to check out the latest news about the ferry crisis. Jings, Crivens and Help ma Boab behind the counter is the guy who played the Bluegrass song Rain Go Away. I go up and ask him about his Dobro guitar and he tells me it's home made from two old guitars and the Dobro sound added to it. Wow I am impressed. I am so impressed I offer to buy him a Rye.

MAGGIE:

Slept late and got up around 10 ish. We went to the visitor centre in Lerwick and lo and behold there was the wee guy who played the bluegrass. We had a nice wee blether.

We had noticed already that Sheila Fleet, the jeweller, has a shop in Kirkwall. Even better we find out that her workshop is at nearby Tankerness and she has a cafe there as well at the Auld Kirk. It would be a dream to actually meet her. She is dear to our hearts as it was Sheila Fleet who designed and made our wedding rings.

Everything Sheila Fleet does is done so well. So much thought goes into it. We have a very, very tasty lunch in the beautiful surroundings of the Auld Kirk. We had asked before lunch if it is possible to meet Sheila and the kind wee lassie at the counter said she will see what she can do. After lunch, we are delighted as she beckons us over to meet Sheila and we tell her about our wedding rings and our wedding in Burleigh Heads. She says it just so happens she is going to give another couple a tour of her workshop and she offers us to join them.

We are thrilled to be shown the process. It is a very inspiring atmosphere. We are shown the intricate details of the work that goes into making such original and stunning jewellery. One of her collections is inspired by the last walk in

Autumn with her late husband who sadly died of pancreatic cancer. She talks fondly of him during the tour. He loved photography and they both inspired each other. She told us that many years ago (she is a youthful looking 73) they asked Scottish Business Enterprise for a second hand camera and the rest is history. What a fantastic and inspiring story and what a place.

We got to meet a young apprentice who was practising making bees. She was also making a ring which was to be polished up. It reminded me of the master/apprenticeship relationship of the Renaissance time of Michelangelo etc. This sort of skill takes years to develop.

Sheila also told us that her next theme is the cosmos. She says she can feel things could be 'in the air' and tells us about the time her husband went to Australia trying to find her opals. Out of all the opal outlets he randomly chose the one that loved Sheila's work and owned some of her jewellery. They became friends and all went out to a Greek Restaurant with opera singers at the table. We spoke of our fossicking experience in Australia.

All in all we had a fantastic time. The wee apprentice offered to take our photos with Sheila. The tour took two hours and then she offered for her wee apprentice to polish our rings. She told Pete he reminded her of a wee Billy Connolly. Pete told her he had actually met Billy in 1973 in a wood…

PETE

We realise that we are not far from the Orkney jeweller Sheila Fleet who made our wedding rings. They are made from Orkney silver or should I say forged? I say forged because these my friends are no ordinary wedding rings. They engraved with real Viking style and real rune symbols which symbolise 'dreams of everlasting love'. We love them and decide to go and see the place where they were made.

The workshop is next to an amazing cafe and showroom exhibiting the beautiful and original jewellery designed by Sheila Fleet. To our surprise, when we ask the girl behind the counter, she confirmed the lady who made them, Sheila Fleet, is actually there. She took time out of her busy day to talk to us. We get to show her our rings and she is delighted to hear our story.

She invites us on a personal tour around her workshop. We get to see the beginning of the process where the design drawings are created right through all the stages to the finished project/item. They are works of art for sure. She even tells us to take our rings off so one of her little elves to clean and polish them for us. The little elf who polished our rings was a real wee cutie. She employs local people. The designs are so intricate. We are so impressed to see the work that goes into a piece of jewellery. It really is art. We are very happy to have met Sheila and her young apprentice and promise to keep in touch if we can. A great time was had by us and our rings are looking brand new and shiny. Just before we go the wee elf kindly offers to take our photos with Sheila. Another dream come true.

Where's Da Orcas

MAGGIE:

I am glad the ferry was cancelled on stormy Saturday night as the journey over on Sunday night was so calm I slept the whole way. When I woke up, we were entering Lerwick harbour. Mind you, if we had gone over on the Saturday, we would have been able to spot the pod of Orcas in Lerwick harbour on Sunday. The photographs looked amazing. We are excited that we have the chance to see orcas and otters. Seemingly Shetland is the best place to spot otters. I love otters, so here's hoping.

Shetland definitely has a Scandinavian feel to it with many of the timber houses painted in pretty pastel shades. We decide to find somewhere to park for tonight out with Lerwick. We take a run down the coast towards one of the few campsites. On the way there, we pulled in to see a herd of Shetland ponies. They were so small they were up to my knees. I think there must be two different sizes as these are the smallest I have ever seen. So cute.

We arrive at Levenwick campsite and fall in love with the wee kitchen seating area. It sits up high looking over the Levenwick village and bay. The view is so lovely and we feel it is a great place to write our book. We find out that there are workmen coming to start upgrading the hall and toilets/showers for next year. It is the end of their camping season in a couple of weeks. It looks like our plans to base ourselves here won't be happening. Hope we can find somewhere inspiring.

PETE:

We could be doing with a rum when we find out that the ferry journey to Shetland actually takes seven hours to get there on the North Atlantic. All that tossing about and rain...sheesh...at least we get to look forward to hopefully

seeing orcas, otters, wild flowers, birds and of course AURORA BOREALIS—
The Northern Lights.

Well, Well, Well. That was a nice surprise. The seven hour sail went very smooth, with calm and clear seas and no one turning green. Sadly no orcas today. Still, it's early days. We hear there is a place not far from here which has a camping facilities so that's where we are going. That's after we stop someone for a jump start as we left our lights on and drained the battery. That's one way to get to know the locals. TIP: always carry jump Leads. Good job we brought them. Welcome to Shetland lol.

We reach Levenwick. It's a lovely place with views out over the rough Atlantic. We will be safe up here…NOT. When it gets dark, a ferocious gale gets up, the van rocks and we hear a bang. I can hardly open the door and when I do get out I see the roof case lid flapping. The wind has blown the catch off. Luckily, no clothes have blown out. We decide to move the van up beside a high wall which helps a bit. I can't help thinking how lucky we were on our seven hour ferry crossing.

MAGGIE:

Crikey it was a very stormy night with gale force winds. Took us by surprise, so lesson learned and from now on I will be checking the weather app every day. We were a bit naive but we learned to shelter behind walls after our roof box hinge clasp got blown off. One of the workmen recommended getting ratchets.

It's turned out a nice day and after a blissful shower we went a walk to the village beach. We saw an otter's paw print and take that as a good omen. The weather changes every five minutes here. Definitely live with the elements. It feels great.

We go for a drive down the coast looking for orcas. We ended up in Bigton which is a small village with a big heart. We go into the Community Shop and the woman serving us says she recognises the accent. Turns out she and her husband lived near Culzean for twenty one years. They retired back home to Shetland a year ago. She recommended going to Ireland which happens to be the small hamlet just over the road as there is camping there.

That evening was very exciting as it was only our second night and a group of people were gathering to orca watch. Seemingly a pod of orcas had been spotted on the other side of St Ninians and there was a good chance they could nip into the bay where we were parked. We were unlucky that night as they passed us by. Later on that week we saw a photograph in the local newspaper taken across the bay at St Ninians that very evening. It was of a guy in a canoe with a huge orca fin next to the boat. Amazing! We were so close to seeing them.

The next morning was sunny and calm. We had noticed a huge pile of wood down the road which had put us in the notion for a fire. We found out from our friendly community shop that the wood was in fact for November 5th and not for Up Helly Aa as we had wrongly guessed.

We decided to have a fire with the bag of wood we had been carrying since July when we had parked at a campsite near Solsbury Hill. We so enjoy a campfire, but it has been too windy. We made two dinners from a pound of mince. One spag. Bol., and the other mince, onions, carrots with potatoes sliced on the top. We had a quiet day and it was bliss.

Speaking of Solsbury Hill, it is Pete's favourite song and it's Peter Gabriel, his favourite singer. It was a dream of Pete's to walk to the top. A dream which has now come true. Please listen to this song if you can. It is inspirational.

PETE:

Next morning we head back to Lerwick and splash out on a pair of ratchet straps. There is no way the roof will move again. It turned out to be a wise buy. While we were back in Lerwick, we decided to go back to the little harbour port between the marvellous Shetland Museum and the Mareel Music, Cinema and Education Centre. We are looking for seals and maybe orcas.

The Museum cafe overlooks the harbour and the coffee is good. It reminds me of that coffee in Orkney in the pub except no cherry liqueur…

All fed, watered and with roof case secured, we drive off to St Ninians beach which is next to a small place called Bigton. It has one shop run by the community. The beach is to die for. I found out it is a tombolo beach which means two beaches join together by a long strip of sand. Oddly enough when we arrive the two beaches had the tide coming in at the same time. It leaves a causeway type of path to walk between the two seas and onto the island of St Ninians. There is only certain times you can cross the sand. If you don't time it right, you are stuck with sheep and birds and if that is not enough the edge of the cliffs has a sheer drop into the Atlantic.

There is an old Monastery and to build it they must have carried the blocks of stone across the causeway which is quite a feat. We managed to get back over to the mainland before the tide came in.

The woman in the community shop recommended that we could camp at the small car park for the night and maybe see some orcas. We find out that the car park is next to a graveyard. No wonder it is so quiet. Still no luck with the orcas even though they were seen nearby.

MAGGIE:

We went a walk to St Ninian across the beautiful tombolo beach where two seas meet. We walked across a very narrow cliff. Good job it's not windy. The beautiful clear water reminded us of New Zealand. In Switzerland, Iceland, New Zealand and Shetland, we have seen first-hand the pure, turquoise water from glaciers. We are very, very lucky indeed. We had a memorable day there. Pete made a peace sign on the beach made from feathers. We will definitely be back here.

When I went to the public toilet, I had a strange feeling come over me when I saw a poster saying I JUST FROZE—FREEZING IS A NATURAL RESPONSE TO TRAUMA. It was as if the words jumped out at me and then touched me deep in my core.

The next morning at the Community Shop we got blethering. It is better than a Tourist Information Centre-very friendly and welcoming. We found out about Sunday teas. Bigton is having one later on in the month. It is fundraising for the fireworks display to which we have been cordially invited. Seemingly Sunday teas are a tradition in Shetland whereby lots of people make soup and cakes etc., and the community comes in droves to support the cause. We are looking forward to sampling some. The first one is tomorrow at Walls except it is on a Friday and is a Church tea.

Sunset in Shetland

PETE:

After our sleep next to the graveyard, we decide to visit a place called Walls pronounced Waas. Our friend in the community shop recommended that we go there especially as it was Friday and there would be a Friday Tea in the Church. Most weeks, mainly on a Sunday, in Shetland a community somewhere will be holding a Sunday Tea. They are delicious and help raise funds for various community events or for the church. She told us we would come out full up and would maybe get to taste the famous Reestit Mutton Soup. That will do for us mate.

The Church is busy and when we find a table and both order the Reestit mutton soup, that's dried mutton to you and me and it's superb. Enough to want more. We learn from one of the ladies that some of the sheep on Shetland eat seaweed which is gives the meat a unique taste. I think my soup may be one of them. True to her word we came out of the Church full of grace or should I say lunch.

The wee village of Walls is very scenic and the more we see the more we think it is a bit like New Zealand. Just like Orkney it is very Hobbit like with lots of small ancient looking buildings dotted about. We are convinced that everything up here were made by hobbit size people like us. The Picts were definitely a small race of people back then. When we were heading up to Orkney and Shetland, we talked about our heritage and hoping for a connection to Vikings, but now I am thinking Picts.

As we move out of Walls we end up camping the night at a small scenic harbour, of which I'm afraid the name escapes me. However, the night sky doesn't escape me. I really thought we were going to see the Aurora Borealis, it was that bright, but sadly no it wasn't them. It was a typical beautiful sunset in Shetland though. The sky was lime coloured, then yellow then red, then all the stars came out. Really amazing.

MAGGIE:

It's another beautiful day and a lovely drive up to Walls (pronounced Waa). Again a lovely wee village and we were made very welcome at the church for lunch. We did a bit of homework before we came up to Shetland. We are keen to taste some of the traditional food and this is a great way to do it.

We order reestit mutton soup together with a beremeal bannock. The reestit mutton is cured dried mutton which adds a unique and distinct flavour to tattie soup. Beremeal is a type of flatbread made from a barley harking back to ancient crofting times which is only made in Orkney, Shetland, Caithness and a few Western Isles. All in all it was delicious combination. I love trying traditional food wherever I'm travelling. We also looking out for rumbledethumps which is seemingly a dish made with cabbage and potatoes a bit like English bubble and squeak. We love the word rumbledethumps so we have decided to try to use it as much as we can, probably as some sort of code language.

I have also noticed lots of people wear the traditional woollen jumpers Shetland is famous for. It is Wool Week soon so that will be interesting. Shetlanders are very proud of their traditions and most of all their language and dialect.

I have noticed that at times to me, people in Orkney and Shetland sound Welsh, which is interesting because I read somewhere that the Welsh language was used in Scotland way back in Pictish times. It feels so comfortable using our Ayrshire dialect here and most people seem to understand us. It frees you up to just be yourself and feel comfortable in your own skin.

We walked around the village. There are lots of derelict old croft houses and lots of planticrubs. I got speaking to a mechanic as we thought he might have a wheel trim. He asked us why do we want to come to Shetland in the winter? It is not the first time we have been asked this question. We mentioned the wildlife, the Northern Lights and Yell to trace family history. He said his wife had said she was treating him to a holiday and he thought Greece maybe but ended up it was Yell. He says Unst is worth a visit but he usually just drives right through Yell and it only takes ten minutes. He wife insists there is more to Yell than meets the eye though. I am not sure he is convinced but I look forward to seeing if she is right.

PETE:

Next morning when driving we notice a sign which says Chinese Night: Friday 20th 8 pm at Community Hall. Nice one I say to Margaret, bet they have loads of Chinese lanterns and dancing dragons. Maybe we should go?

When we arrive in a village called Brae, we got talking to a biker girl who likes Deep Purple. Not only does she tell us of a nice place to go camping, but she tells us that sadly the Chinese night is not in fact a festival of some sort. It is where Chinese takeaway food gets served in Community Halls on certain dates. This saves the locals travelling quite a distance to get a takeaway. No surprise that it is always popular. Ah well, it's a bit windy for Chinese Lanterns anyway.

We decide to take our biker friend's advice and go to Eshaness. When we get there, we discover, yet again, just how stunning Shetland and our own country can be. It's another golden coloured beach and you can't help thinking that you are definitely going to come back again one day.

After camping the night, we head to the beach to see if we can see some dratsies, that's otters to you and me. We have been here long enough to keep a wary eye on the tides and the weather here. You can get caught out quickly.

The beach is deserted and looking for things on the beach is a good pastime. Sadly, there is always that monster: PLASTIC. Margaret hates it with a venom. When I look up, she's actually at the other end picking it up… It reminds me of a time on Ailsa Craig when I rescued two gannets whose bills were cojoined because of plastic…

As Margaret carries the plastic up into the field there seems to be some sort of pile there, so we added to it assuming that the council or the farmer collects it and takes it away.

We never saw any dratsies or orcas. There is a pod which goes round the whole Island, but we have yet to see them. In fact, we just seem to be missing them all the time…we have one more look and then it's time to head back down towards Vindin. It's Sunday and there is a community lunch to sample and also we heard there is a car boot sale on. Another nice pastime I like. Here's hoping we get a bargain.

MAGGIE:

It was good to have a laugh with the mechanic and now it's time to move on and find a place to camp for the night. We find a lovely wee harbour up towards Muckle Roe. It has a long 'jetty' boardwalk. Beautiful to look at but not to walk on as it was old and rickety. We got to see another wonderful Shetland sunset.

Boardwalks hold a place in our hearts as Pete had a lovely picture of one and to us they symbolise taking a leap of faith in order to move on to bigger and better things. Even if you don't know what is going to happen, let's face it, the unknown is scary, you know in your heart if it feels right…just face the fear and metaphorically jump.

We head to Brae and stop in at the Co-op. It is Gold Cup day at Ayr Races and we were looking for a Bookies. As we asked around a lovely lady appeared and asked if she could help. Turns out her dad was the last Lighthouse Keeper on the Isle of Unst. She told us about a great campsite at Eshaness and that she reckons it is the nicest beach in Shetland with its stunning cliffs and rock stacks. She is a biker and goes on a rally every Sunday and usually ends up where there is a community tea. It's so helpful to get local tips and advice.

We head up to Eshaness. She says there is a lovely cafe at the campsite. It is a great run up and we pass some gnomes and also a gang of multi coloured trolls which were left by a family doing a charity road trip. Guess what on the way up? We find a wheel trim hanging on a gate and it fits!

The cliffs at Eshaness are spectacular as are the Drognes stacks. The cafe is an ideal spot to sit and soak up the views. Pete came across a book of spectacular photographs and history of Shetland written a while back by a French guy (the waitress serving us happens to be French) I notice in the book that the birds here are all called by different names to the mainland such as Bonxie for Great Skua, Muckle Scarf for Cormorants, Dunter for Eider Duck and Tammy Norrie for Puffins to name a few.

The campsite is simple and yet stunning. The showers are the best yet with lots of room. We parked next to the tiniest caravan I have ever seen. Next morning we go for a walk on the beach and I made the mistake of spontaneously picking up plastic which ends up being a bit of mission. I think it is better to organise a community pickup and hopefully get people thinking about consequences of actions.

We saw a seal right up at the shoreline, riding the waves. It seems to be really interested in us. Most days we see a solitary seal looking over at us.

I quite fancy going to a gospel church in Shetland to see what it is like. Pete tries to explain what a gospel church is in Scotland. There will be no singing he says-it literally means sitting reading the gospels. Indeed a church member explains that to me too. I politely decline as it sounds dull and not uplifting at all.

As we drive off the wee guy gave me a big smile and a wave and told us we are welcome anytime which I thought was sweet. I feel that people sit and read and read the bible, all coming to different conclusions. There is only one line that I personally think that everyone needs to read…the word became flesh and made his Dwelling among us… I just don't think many people realise the power of the words they speak and think.

PETE:

Well, that was different. Nothing much to report on the car boot sale, except that a woman (who looked a wee bit dubious to me anyway) saw me picking up a book at her stall. I already have the book as it belonged to my late partner and it was something to do about angels and the cosmos. I recognised the cover. I was just about to say to Margaret look I didn't expect to see this book way up here. I've got it in a box somewhere. All of a sudden the woman comes running over to me.

Hey she shouts do you want that? It's a good book.

I've already got it I say

It's yours if you want it

I have it I repeat

Half price she says shoving it in my hands.

But, I've got it…

Are you taking it? She says

No it's ok. I calmly tell her that I wasn't pointing at it because I wanted it, but in fact to tell my wife that I was surprised to see the same one way up here.

There's more over here she says pointing at lots of boxes.

By this time, the other stall holders are all a bit embarrassed by her antics. With that we left quite promptly, but not before we had our Sunday tea.

We are taking the chance while on this peninsula to go and visit the oldest working Church on the Islands. It's at a place called Lunna. It's very old and atmospheric, overlooks the sea with gravestones as old as the hills. We put something in the collection box and move on.

Sadly there are still no orcas. As we drive along I keep looking at the rocks and coastline. It's beautiful and very dangerous. Someone told us at Eshaness beach—one day there were a group of French tourists. They were admiring the lovely beach, but sadly didn't study the tides properly. They ended up having to be rescued off some rocks by the coastguards. It must have been quite an ordeal for them. It is wise to be careful, not only at Eshaness, but everywhere on the coastline.

MAGGIE:

We are going for Sunday tea at Vindin. We arrive an hour early and walk round the car boot. The people are not as friendly here as elsewhere on the Islands.

After the tea (there was not a lot of traditional food on offer), we nip into the local shop before driving down to Lunna which has the oldest kirk still in use (1753).

I noticed a shop name and I recognised it from the photographs in the book at Eshaness. I loved the fact that son of Hercules is an actual name. Yes, I have seen Williamson, Johnson, Jamieson but to see Herculson now that felt like right ancient Viking and made my day.

The auld kirk was interesting to see and reading the inscriptions I feel so lucky to be living in this era. I have done my time. The Laird's house stands on the hill overlooking the harbour and seemingly this was not only to keep an eye out for smugglers, but also to keep an eye on his workers and tenants.

As we head back down the road we pass a War Museum in the middle of nowhere. It is a personal war collection which has been made public and run by volunteers. I can tell it has been a real labour of love and when I see the photograph of the guy who had the collection I could see the family resemblance to the guy who was volunteering. Sure enough, when I asked, turns out it was his uncle. I learned quite a bit about the amount of Shetlanders who fought and died in both World Wars.

Vegetarian Butcher

PETE:

We are heading back to Lerwick, Shetland's capital, where they've got the best library you will probably see in the UK. I can't help thinking about the Army and Navy Museum we happened to come across in an old corrugated iron shed. It's run by volunteers. They've got all the old motorcycles of the day complete with goggles; a mannequin of a young woman dressed in the fashion of the time; old army and navy uniforms; artefacts and lists of names of all who served and all who never came back. We spent quite a bit of time there wandering around. You keep forgetting and not realising the impact Shetland had protecting Scotland during the War.

They have a good sense of humour up here as we discovered when we passed a sign warning of squirrels (a picture of a squirrel) and of course there are no squirrels, but believe it or not there are gnomes and trowlies (that's trolls). We saw them earlier right in front of us, sitting on a rock laughing at us. They weren't of course, at least I don't think they were.

We decide to camp at the harbour in Lerwick in between the big Museum and the Picture House come cafe bar. It's not far from the library either. After we had our breakfast, we went walkabout and discovered a butcher selling local produce. As we go in Margaret gets talking to the lady behind the counter and we discover that although she has worked in the butcher for years she was vegetarian for years, but now she has started eating meat again and she says she feels better for it. Probably the extra protein. A vegetarian butcher who would have thought that.

MAGGIE:

We end up at the harbour in Lerwick. It is a wonderful place to stay. It has a welcoming feel about it. The Mareel Arts centre is fabulous and the arts are supported very well up here. Lots of musicians, painters, sculptors, writers and of course, knitters. We see a poster advertising an artist and her exhibition is called Stoal which sounds interesting. It is at Scalloway and we haven't been there yet.

We come to the Lerwick Library which is in a former kirk. When we go in, there are beautiful stain glass windows and it is just stunning space to sit and read or write. Pete found a Robert Plant biography and a Van Morrison biography which brought to mind an amazing concert we went to in Dublin where both Robert Plant and Van Morrison played for nearly four hours between them. That was a dream come true for Pete and he deserved it as he had recently passed his driving test at the age of sixty three.

I got a Shetland dictionary and a couple of books about the old Nord language specific to Shetland. I love libraries. I worked in one for years and I loved it. Lots of interesting stuff coming up including a Billy Connolly film of a recent show. We both fancy going to that, but think we will be on Yell when it is showing. Fingers crossed it will be showing there.

Next day we decide to go to the Shetland Museum and spent a couple of hours going round it and only got to year 1700. I said to Pete, I need a rest or I won't take any more information in. We decided to come back another day and do part two. It is free admission. I am so proud to be Scottish. Free admission to museums so anyone can visit and beautiful libraries. You can feel the sense of pride here and it frees you up to be yourself and to think outside the box. I really do feel that is why we invented so many wonderful things. We believed in education for all people long before most countries. It freed people from the shackles of control from the churches of the time.

PETE:

On our second day in Lerwick, we can't decide whether to go to the Library or the Shetland Museum so we toss a coin. Margaret decides it's the Museum although I didn't see any coin getting tossed. We are planning on going to the Islands of Yell and Unst.

Inside the museum we discover just how large it is. It's got artefacts and memorabilia of all kinds. Sections on sailors and sailing, Vikings, farming, Shetland knitting. There are loads of photographs of housewives carrying baskets of herring on their head whilst walking and knitting at the same time. I think that is termed multi-tasking. They were definitely a hardy bunch. The museum offers so much that Margaret has to sit down as it is all too much to take in all at once. We decide to come back another day.

We also joined the Library as it covers all Shetland's islands and it's free. It is a huge building with stained glass windows and very friendly staff. I look round and discover a book on Robert Plant that'll do for me and a book on Tolkien's early life that'll also do for me. Margaret got some books too and so we decide to go to the pub/cafe at the Mareel later on to read.

Be Afraid, Be Very Afraid

First though we decide to go to the Heritage Centre. Margaret is convinced that she has Viking blood in her and I've always had a kinda Viking/Pictish feel about myself. So off we go to try find out about our heritage. I was telling Margaret about my Granny on my dad's side. When I hear the accent here, it reminds me of her. She was an amazing woman. She only had one arm and was actually in the Sunday Post because she had her operation to remove her arm without anaesthetic.

The Town Hall where the Registry Office is located is awesome. It is huge, so before meeting the Registrar we take the chance to have a wander. There are lots of things on the walls. We head upstairs to the large Hall and it's magnificent with huge stained glass windows all around. They each have an important historical figure on them and are so colourful. There's one of a giant Viking called Ragnar of the Hairybreeks. Only kidding, but would love it if there was though.

I learn I am descended from Bilbo Baggins…obviously I'm joking. However it turns out my dad was born out of wedlock and was three years old when his mother married Andrew Burleigh. I found out that she married Andrew Burleigh on 14th February 1919 St Valentine's Day 100 years before me and Margaret. No wonder they say who do you think you are? Be afraid, be very afraid. It turns out there is no trace of my dad on record under Andrew Burleigh. No Birth Certificate. So where did he come from?

We have to come back next week while the Registrar checks the maiden name of my gran, as it is unclear. We will check more closely later as my cousin's husband is a minister and he might have access to Parish records and he once said to me that there was some 'skulduggery' in our family. Looks like he's not kidding.

MAGGIE:

We go to the Town Hall which is such an impressive building and it was built by the people of Lerwick. Lots of beautiful stained glass windows. We are going to the Registrar to try and find out about Pete's granny as he thinks she may have been born in this area just remembering her accent.

Pete knew her as Granny Burleigh but doesn't know her maiden name. The Registrar tries to find out by looking up his dad's date of birth but there is no Andrew Burleigh born on that day within ten years each side of the date of birth. Pete knows her husband was also called Andrew Burleigh and she finds their Marriage Certificate. We are surprised to find out they were married on 14th February 1919 exactly 100 years before our own wedding day. Her maiden name looks like Rush but it's difficult to make out. The Registrar is going to contact Edinburgh to find out the maiden name and we have to go back next week.

Crikey we feel like detectives. It was the days of the first World War and it looks like Pete's dad was not born Andrew Burleigh but perhaps his biological father was killed in the War and he took on the adopted name of the man who married his mum on St Valentine's Day 1919. I like to think it was a marriage full of love.

PETE:

Moving on we come to a place called Scalloway where the sheep are very woolly and bigger than the ponies. We stop to fill our water bottle at a fountain that has a lion's face and a tongue for a spout. It reminded me of the Chalice Well fountain head at Glastonbury. We say cheerio and thanks to the big hearted lion.

We are heading back to Levenwick and once there I go for a wander. To my surprise I find the catch that got blown off the roof case. Imagine that. After that storm. Even better, I fix it back on so with two ratchet straps and a locking catch so our roof case is going nowhere.

In the kitchen, we meet a young German girl travelling on her own. She admires my Lord of the Rings ring round my neck. I show her the photo of me and Margaret standing at Bilbo's door at Hobbiton, New Zealand and she loves it. She then tells us that she has read all of Lord of the Rings in English. I think that's amazing. She is German and reads the English version.

She tells us that on Unst there is a full size Viking Longship and that when all the tourists left she slept there for the night. We wish her well as she heads on her travels. We go to bed so that we can get up nice and early to visit Sumburgh.

MAGGIE:

Decided to move on as we want to go to Scalloway to see the Visual Poetry Exhibition called Stoal. Very atmospheric seeing and hearing the poems in visual installations. She used Shetland dialect which has Icelandic connotations and used mythical other worldly creatures and figures to represent the elements. The wind and the sea are such an integral part of Shetland life.

There was one visual where her Granny was talking about occurrences where apparitions visited. One girl she knew saw her fisherman boyfriend and thought he was home. It turned out she found out later that he had perished in a storm and he was still at sea at the time she 'saw' him.

Scalloway is a lovely picturesque wee place with a variety of bright coloured houses. We went into a local cafe/bar for a couple of Irish Coffees to heat up.

We find out the next morning that the Orca pod were in at the harbour at the very spot we had camped the previous three nights. Gutted. That is three times we have actually just missed them. We head to Levenwick campsite to do the laundry.

Once there we met a young nineteen year old German girl who astounded me when she told us she had read Lord of the Rings in English. I struggled to read it in English and have no chance in German. She is travelling alone in Scotland, paying for her travels by working and getting room and board in return. She has slept in a few waiting rooms too.

We watched Metropolis on DVD. Pete bought it in a Charity Shop in Orkney. It really is a classic. Up there with Animal Farm and well ahead of its time. Some would say it is Science Fiction's first cinema masterpiece. It was made in

Germany in 1927 and when I read the blurb description it says it's about a beautiful and cultured utopia existing through the miserable, mistreated workers in the underworld and how they try to rebel. Sounds familiar.

Something told me that Hitler would have liked it and sure enough when I checked the internet he seemingly loved it and tried to get the Director Fritz Lang to be part of his propaganda machine. Goebbels succeeded with Lang's then wife who was the actual screenplay writer. She stayed in Germany whilst Fritz Lang left pronto.

We took no chances and got permission to park at the front of the car park and had a restful sleep.

PETE:

We wake up nice and early, get tea/coffee and breakfast. We jump into Gabie the van ready to head to Sumburgh Airport to see what it looks like as it is not far from the beach. Only it won't be today. Gabie won't start as we've left the lights on all night and drained the battery. There is nobody else here and a sign says the proprietor comes around seven or eight o'clock tonight. It's only ten in the morning so we go for a walk half afraid we miss any cars coming so we don't go far. We end up sitting looking at the sea for orcas and bird spotting.

Finally a few cars start to arrive. It's dark now and we find out they are all choir members who happen to use the community hall for a sing song. Yeehaa. I say hello, explain the situation and that I have a pair of jump leads. One of the women swings her car round with the headlights on and all together we start singing 'Start me Up' in my head. We pray the jump leads are on correctly when our Guardian Angel starts the engine and to our delight we get Gabie on the road again. Man, I am so glad I remembered jump leads.

We find a parking spot at a beach near Sumburgh and make sure the lights are off before going to bed. It's funny how doing nothing can make you tired.

MAGGIE:

We left late to get to Sumburgh and ended up at the only Hotel in that area. All the shops are shut and there are no big supermarkets down this way. It's like stepping back in time. I like it. The Hotel is really nice. We have a plate of mushroom soup and a strange foosty tasting local ale called Skatness from Valhalla Brewery, Haroldswick. Can't say I enjoyed it. I was not surprised when I heard that it has gone out of business.

We end up at the 2019 Scottish beach of the year and wake up to a beautiful morning. We decide to drive back to the Hotel and walk to the Lighthouse. It's a great walk along a craggy coastal path. There are lots of twitchers going around with enormous camera lenses. Seemingly they are looking for a rare Sandpiper and a Little Bunting from Siberia both of which have been spotted. On the way, a guy gets talking to us and it turns out its local wildlife explorer.

He shows us a photo of a Little Bunting which has been taken at the Lighthouse whilst we were in having a coffee and admiring the view.

We have been so lucky with the weather. We got talking to a couple in the cafe who moved to Shetland from the Hebrides and they much prefer it here. The West coast has much more rain. She asked someone to add us to the WhatsApp group for spotting orcas, whales and other wildlife. We still live in hope of seeing something exciting in nature. Back at the Hotel we have a Guinness and then back to the same beach tonight. It is a nice quiet spot.

The I Just Froze poster last week has had a big effect on me. I have come to realise that I suffered PTSD as a child and it's true to say that it feels like my emotions were suspended in a sort of frozen wilderness. I am determined to master my emotions and put my hurt child to bed. I know what it is now and it does not scare me. I used to be scared of my emotions and with no one to listen to me they swirled around inside my head. I have had a lot of unexpressed emotions. The secret is to acknowledge them and feel them and then they will dissolve. The past couple of days I am letting go anger from my past. I can only describe it as like having a forest fire blazing. It's like a million nettle stings. The old energy is burning away so that my new growth can come through. I am finding my way to love through my heart, a higher love and a higher state of consciousness.

PETE:

When we wake up, we drive back to the large Hotel. There is a helicopter sitting on the grass. We park Gabie and go for a walk up the cliffs to the lighthouse which has a cafe overlooking the cliff tops out to the sea. On the road up, we stick to a well-worn path next to the guard wall as it is a sheer drop to the Atlantic on the other side.

Once we get to the top we see a lot of people with cameras, tripods and expensive looking long lenses. At first, we wonder if they have spotted orcas so we hurry up only to discover that they are actually twitchers—people who follow birds. They have been known to travel across the Atlantic for anything up to twelve hours to get here in the hope of spotting some rare birds often blown off course. We find out that they are looking for and hoping to see a Little Bunting from Siberia which has stopped off here in Shetland. One of the men walked up to me and asked if I'd put some eye drops in his eyes for him as he needs to do it four times a day. I duly oblige and also tell him to eat plenty of blaeberries (bilberries) which are good for the eyesight. After that we head for a coffee.

A woman shows us a site that follows orcas and as we're dying to see them we join. As we leave we speak to a guy who asks us if we want to see photos of the Little Bunting and to our surprise it turns out he is the guy who runs the orca site. We are well chuffed and as we walk away and move down the hill, there right in front of us is an orca. We run down and get our photos taken standing next to it. Margaret stands clapping it's head and as it's made of iron, I climb up on it and get another photo.

We climb over the dyke and through the gate back onto the steep path back to the helicopter hotel. As we are walking down just the two of us, Margaret stops me in our path. Right in front of us is a little bird and it's putting on a show, bouncing about and chirping away quite the thing. It lets us follow it for ages. As Margaret loves birds she really is in her glory. Her wee face is shining with joy. Man, it's great how the simple things make your day.

MAGGIE:

We wake up to another nice day. There are more twitchers around looking for the rare Sandpiper. We go for a walk along the beach. A family were building sandcastles which brought back great memories of those kind of days spent with my own two girls. The beautiful water reminds me of New Zealand 's melting glacier water. So pure.

The airport is nearby and the runway goes across the main road which is a bit surreal. It's time to head up to Bigton Community Sunday tea which is fundraising for the fireworks display. We have been invited and have been made to feel very welcome by this small community. It is very busy. The food is scrumptious starting with a selection of soup. We chose the Cullen Skink and then delicious tray bakes especially the one with lemon curd filling.

We saw the friendly lady who stayed in Ayrshire and she introduced us to her husband and daughter. We were invited to come to the November fireworks display so we mark it in our diary. On the way out, Pete bought some raffle tickets.

We pop into Levenwick on the way over to Lerwick. It's really quite close to Bigton. We then visit Sandwick and Pete notices that adjoining Holswick has a very nice Visitor Centre. We had a lovely chat with a lady there. We were discussing Pete finding out his dad was not born Andrew Burleigh. She told us that there is loads of family history going on and that 'cousins' turn up all the time. She also tells us about Wool Week which is this week. We can see that there are folk in the Visitor Centre knitting and spinning yarns in more ways than one.

We end up at Lerwick at our usual spot.

PETE:
We are going to Lerwick again, but first of all we do two things. The first is to stop at Levenwick to do a laundry, shower and dishes and then head for the Sunday tea at Bigton. Only one word—delicious. I buy raffle tickets as I head out the door.

We are in Lerwick again and we are going to visit a place called Clickimin Broch. I noticed a sign as we arrived back at Lerwick. I am very much looking forward to visiting it since it's based on an old Pictish settlement.

It lives up to our expectations. The board shows you what it would have looked like, in its day, hundreds of years ago. It's just a wee bit off the road and very atmospheric. In fact, it has got a really strange and weird vibe to it. I can feel it. We both can. I decide to take some photos. I manage to take three then my phone zaps out. Completely dead. This is strange because I know for a fact that there was seventy per cent charge. I saw it when I turned my phone on to take the photos. I then asked Margaret to borrow her phone to take some photos and she checks and says she has over sixty per cent battery. This time I take one photo and Margaret's phone gets zapped as well. Stone dead. We ask each other what we think is going on. We decide to go back and recharge both our phones and come back tonight. That's just what we do.

With one hundred per cent charge on each of our phones, we head back to the Broch. It looks great lit up. Very impressive. We walk in lowering our heads through the entrance into the middle…the Broch is round and has three tiers on it. Each one has a purpose i.e. bedrooms, living rooms etc. The ground floor is where the fire is…everyone would be sitting round it…it's very easy to imagine it…so I take the phone out to take a photo. I have kept it off at one hundred per cent so that we have a full battery. I turn on the phone and it turns itself off? ZAPPED. All the energy has gone out of it. I get Margaret's phone and then, after four photos, it goes off as well…This place is full of an energy you can feel and it's so strong that it drains our phone batteries nearly instantly. Yet, it is not scary in any way. It's a cool vibe here and if you are ever in Shetland I'd recommend a visit. We ended up going back a few times. We even took a picnic.

MAGGIE;

Woolly hats and jumpers are everywhere. After all, it's Wool Week. As we drive in Lerwick, Pete feels immediately that we should visit a nearby Broch.

We also went into the Natural Heritage Centre to talk about orcas as Pete thinks he found a whale tooth on the beach. We had a great conversation there about orcas, otters, seals and the fact that New Zealand flatworms are eating our precious earthworms.

The lassie is from Yell originally and so gives us some good advice places of interest to go and visit, especially to see otters. She is going to keep in touch with us on WhatsApp. We realise how lucky we are getting the opportunity to see these places. The wildlife is such an important part of it. We find out that Pete's whale tooth is probably plastic smoothed out by the sea. Aw well.

When we visited The Clickimin Broch, I could immediately feel an amazing energy and felt really at peace. I didn't want to leave. I could have sat there all day. Since our phone batteries were both drained we decided to come back later in the evening and it was just as amazing at night.

It was lovely to see it lit up. I felt blissful. It was such a beautiful feeling. I said a prayer of thanks to our ancestors and also for a friend of Pete's daughter Liz. We were so sad to hear he had just passed away. Rest in peace you were a good wee soul and thanks for being so nice to my dogs.

Next morning, first of all I have a shower at the amazing public facilities which are open all year in Lerwick. Then I go back to the Shetland Museum with fresh eyes. It was very interesting. There is a huge sea turtle which seemingly washed up dead on the shore in 2005. They put it in the freezer to preserve it and then took a cast of it. It made me emotional as I thought about their fragile ecosystem and the privilege of seeing them alive and well, in all their glory, in Australia.

I also enquired about the Yule celebrations which I read used to take place on 6th January (which ties in with epiphany in Spain and other countries around the globe) It was confirmed that Unst and Fetlar still celebrate it. This intrigued me and I would love to experience it one day, along with Up Helly Aa which takes place later on in the month of January.

PETE:

My phone rings and it's the raffle folk to tell me I've won a prize. They have left it in the local community shop in Bigton for me to collect. It's a Nature Journal which is appropriate with all the amazing wildlife we have been seeing. I say to Margaret to have some time to herself at the Museum or shops as I have decided I will use my bus pass to go to Bigton and back.

We say Mmwaah with invisible kisses and Margaret's got breakfast coffee planned already. She is a bit of a Hobbit when it comes to food is our Maggie. At least four dinners a day if she got away with it.

So I go for the bus at the Viking Bus Station with a picture of a huge Viking on the walls. There is a main bus and a rural bus. I wait for the rural one and it's not long till it comes.

There is only me and a woman bus driver on it. I suppose that's why they call it rural. Anyway, since it's just the two of us it's not long until we get talking. I tell her we are doing our family history heritage in the hope we're Vikings and her first reaction was "Oh dear…be afraid, be very afraid." She goes on to tell me that she did hers at school in a classroom experiment, during her last week before leaving school. To her horror here on Shetland she was related to all but one in her class of eleven people and she had been out with six of them. So if you are doing your ancestry: BE AFRAID, BE VERY AFRAID.

When I reach Bigton, I duly receive my Nature Journal and because I have a bit of a wait till the next bus I go a walk back down to St Ninians tombolo beach. That's the place, if you remember, where you can walk across the sand to the adjoining small Island with the tide coming in on both sides of the beach. It's quite odd looking when you see it.

As I look over the sea I wonder where the orcas are today. The time goes past quickly and I realise I better head back up for this bus. I pass the shop again and remember that we have been invited to come back for the Bonfire Night Celebrations on Saturday 9th November. She said there will be a huge fire, fireworks display, food and hot drinks. Then beer up at the hall afterwards. Sounds good to me…

I have missed the local bus so that means there is not another one until tomorrow. *A bit like Australia*, I thought. Over there a woman told us, "Hell yeah, we've got a great bus service here: Tuesday and Thursday and once a day at the weekend." As I walk along I see a couple of big birds. Buzzards I think. After a few miles, I reach the bus stop up at the main road and my phone signal comes back on. It's a WhatsApp from Margaret. She has also been trying to phone me to tell me there is a pod of orcas at St Ninians beach. I've just missed them again! Aw well maybe we will get to see them on the ferry to Yell in the morning.

MAGGIE:

Pete came back with his raffle prize. It's a pretty Nature Journal which is very appropriate considering the amount of joy we have had living so close to nature. We have seen so many amazing sights and creatures. We hope to be making a few entries once we get to Yell tomorrow, especially otters. Pete literally just missed the pod of orcas at St Ninians beach. If his phone had a signal, he would have seen them. Just not meant to be at the moment. So close.

We go into the Mareel to charge our phones and write our diaries. Who do we meet but our Scottish Heritage friend who happens to work here as well. We chat again about wildlife and Pete shows her the Nature Journal he won in the raffle. The first sighting he puts in his Journal is a mermaid called Maggie lol.

We go into the museum as we heard there are some youngsters playing traditional music at a Wool Week gathering. We were kindly told we could sit amongst the knitters, most of whom were knitting whilst listening.

There was a map on the wall with pins to show where the folk, mostly women, are from. They come from all over the world. Mostly Commonwealth countries like Canada and New Zealand but also Holland, Scandinavia and even Asia. I admit to feeling jumper envy and hat envy especially as my Shetland jumper lookalike I have on is 95% acrylic. I feel a bit of a fraud as I can't knit to save myself. It is a nice cosy atmosphere and it's great to hear these very talented youngsters keeping the traditional music alive with their fiddles, banjo and accordion.

Led Zep

PETE:

Before we head off tomorrow, we go back to the fantastic library to hand in and renew our books. We are allowed to take books back to Yell so I am taking the young Tolkien and the Robert Plant biographies.

I loved Led Zeppelin as a young rebel kid in the late 60s/early 70s. It was Robert Plant's voice that I liked and I still follow him as a solo performer. I knew just about everything about Led Zeppelin and I was looking forward to reading the Robert Plant biography. What happened next was just another weird incident of this journey. We went to the Mareel to read our library books. As I settled down to read, opening the book, something fell out the pages. Holy Shit. There on my leg was a plane ticket. A ticket for Sumburgh Airport way up here in the Shetlands, with the name Peter Grant (Led Zeppelin Manager's name) on it. It fell out of a Robert Plant book and Peter Grant died on 21/11/1995. Where on earth did that come from?

While we were in the library, we were told to look out for Bob's Bus Shelter on Unst. It's famous tourist attraction and you can look it up on the internet. Bob got so bored while waiting on the rural bus he put a seat in the shelter. He then went on to decorate it and he changes the theme every now and then. It's had a Hippie Theme, a Goth Theme and just now it's been painted purple and has a Science theme. We make a note to look out for it.

As it's our last night we are going to visit the Museum as they are holding a concert of some sort. The reason for the concert is to celebrate Wool Week. We have arrived in time for the Annual Shetland Wool Week with different themes and events showing different knitting and spinning skills. I am quite happy to give a pound for a badge which simply says Shetland Wool Week. I get the feeling the Islanders like this and in fact I wear the badge long after Wool Week. It's only when we get back to mainland Scotland that I actually take it off and realise that I've had it on so long it had rusted round the edges.

Once in the main hall there are sofas and chairs all taken by women and men sitting knitting, some of them even using their spinning wheels. Most have got a woollen hat or tammy which they have made themselves. All and I mean ALL of them have got traditional woollen jumpers. They are made from the famous Shetland wool. The garments are extremely warm and expertly made. I notice some of the older women are wearing leather belts round the outside of their clothes. I can only think it is to hold their wool or something.

We manage to find a couple of chairs and sit down amongst them all. They are going nowhere in a hurry let me tell you, they are so laid back just enjoying sitting knitting. A man comes on and announces the band are going to play and on walk these guys obviously still at school. There's two fiddles, a banjo, a guitar and an accordion. An older guy is on keyboard which doesn't get played on every song. They go straight into a tune, there's no singing, and within minutes my foot is tapping involuntarily, then my fingers are drumming on my chair. It's a good job there is no alcohol is all I can say.

Whilst this is going on the knitters and spinners are happily doing their stuff as if they were sitting in the house listening to the radio. It is a very laid back feel and you can see and hear that the boys are experts already. They love trying to outdo each other on their instruments for speed and accuracy. When they finished the first tune, I wanted to stand up and clap, but everyone else carried on knitting. Some clapped in approval, but only for a second before they dropped a stitch. Others hit their needles together and some would just nod. There were no whoop whoops, yeahs or fingers in mouth whistling here. It's so informal and laid back. The band are great and if they stay together I reckon they could do alright for themselves.

Well, we have still not seen any orcas or otters.

MAGGIE:

Me and Pete had an amazing talk this morning. One of the things Pete told me was about the time in the 1970s when he and his pals joined a youth club. It was run by a policeman from outwith the village. He suggested they go a trip to Austria and if the boys gave him money every week they could save up to go. It made me cry when he told me what happened. The guy did a runner with the money the week before they thought they were going. I couldn't do that to anybody.

The boys were all looking forward to it so much. Shame on that policeman who knew in those days no one would be interested to listen to a bunch of boys. There would be no investigation, no justice or comebacks for allegations of corruption. Pete did come across the guy again a few years later still working as a policeman, but that's another story...

The pendulum always seems to swing too far and now it's the criminals who seem to have no comebacks. There seems to be no justice for victims of crime and the police seem to have little power or else they just can't be bothered. Hopefully one day it will stop swinging from left to right and right to left and will find balance and calm in the middle.

We got two tickets to go and see Billy Connolly in the Mareel cinema on 19th October so we will make a trip back to Lerwick then.

Before we head up, we go to the Library. It's the best library I have ever been in, except for Drongan, of course, where I worked for fifteen years. I get a couple of Pamela Stephenson books. The biography she wrote about her husband and a book about her sea adventures exploring the South Pacific in the same route as Robert Louis Stevenson and his wife took many years before. The book is really about his wife Fanny Stevenson. It's beginning to feel like Billy Connolly is always somewhere with us on our travels.

We also nip over to the splendid Town Hall to find out if the Registrar has any more information on Pete's dad and granny. She confirms that his Granny's surname was indeed Hush and she came from Berwick on Tweed at the north east tip of England. There is no entry for his dad's birth under the name of Andrew Burleigh. At least, we have a name to try and solve the mystery. Hush.

We also go for one last visit to the Broch before leaving for Yell and then Unst. The energy is still amazing and the phones still drain within five minutes. The people of Shetland are so lucky to have this and it is open 24/7 to the public. I can't help thinking if it was Ayrshire it would be vandalised.

The Picts seem to have been clever, peace loving wee folk and I give thanks for this sacred space. For now, it's up to the north of the Island to catch the ferry and off to Yell we go.

PETE:
WELCOME TO YELL…THE FRIENDLY ISLE that's the first thing you see when coming off the small ferry. It's on a huge billboard with a Viking boat and some comical drawings on it. The Island is only ten miles long so we decide to stay the night in trusty ol' Gabie the van and drive up to Unst tomorrow. Before we left yesterday, we booked two tickets for the Picture House in Lerwick to see the only showing of my pal Billy Connolly's show The Sex Life of Bandages. I say 'pal' because I met him years ago around 1973 at an open air Gala Day in Kilsyth, He was going to present the prize for the fishing

competition. We both stepped out from behind a tree at the same time after nature called. We got talking for what was easily forty five minutes. I've got a funny story about what happened during our talk which if I ever get the chance I'll tell you.

As we drive off the ferry port and head up the road towards the top of the island it's very dark and the headlights pick up sheep at the edge of the road. It's quite spooky and I wonder if there are any ghosts? We turn and park for the night at what must be the middle of the island as it's called Mid Yell. We are thinking this might be the main village here and park in the car park near the Sports Centre. We find the islanders don't mind if you respect the environment. They definitely encourage campers although most folk think we are daft coming up here in the Winter.

We wake up after a good night's sleep. When I open the door, it's a nice morning and we are next to the Leisure Centre and also the School. As we have our breakfast no one gives us a second look. I'm ready for the toilet so I walk into the leisure centre. I am made very welcome to use the toilets. I ask a guy if there is a shop and he says Aye at the top of the hill, you'll see the sign. With that we drive off heading towards the Island of Unst where we get another small ferry, but sadly, STILL NO ORCAS.

MAGGIE:

We get the ferry to Yell at 7 pm. It is a fifteen minute crossing and we were the only folk that actually got out the van. No sooner were we out when we had to get back in. We had arrived at Yell. Finally, after getting a bit lost, we end up at Mid Yell for the night.

Next morning we nipped into Yell public library and the librarian was very friendly and helpful. She comes from Unst and gave us lots of good tips. She also told us that there are always free council houses and jobs on the island as a lot of people come and go because it is so remote. She was saying that a lot of the islanders have two or three jobs through choice. I had already noticed that with our friend the Scottish Heritage lassie. She tells us her husband works in the fish processing factory Monday to Friday, then he drives the local bus at the weekends plus he is a leisure attendant and she is a library assistant and a disability assistant. I suppose variety is the spice of life.

We got talking about family history and she said she had her DNA tested. There had been a rumour in the family that an ancestor had married a Native

American. She was disappointed this didn't show up but it came through that she had Norwegian blood. I pointed out that it could be Sami blood which are the native people of the north of Norway. Some things just can't be proved.

After saying our cheerios, we manage to get the ferry and it took seven minutes to cross to Unst. We drive up to Haroldswick which is the top of the island and approximately seven miles away. We stopped at the Viking Longboat and replica Longhouse just before Haroldswick and end up camping there for the night.

I'm Spartacus

PETE:

We drive off the small ferry which took a very short time for the crossing. Most people stay in their vehicle, but not me. It's a sailors life for me and I go up on deck to spot orcas. I don't see any. They keep eluding us. Who knows when we will see them or otters. We keep telling ourselves to keep the faith. We did that in Australia and saw a duck billed platypus on our last day in Burleigh Heads. We did it in New Zealand and saw a kiwi again nearly at the end of the trip. We keep the faith and drive on as there is more to Shetland than orcas and whales.

Unst isn't that long. In fact, it is smaller than Yell. As we drive along we realise just how remote we are and then totally out of the blue we come across a Viking Longboat and next to it a Viking Longhouse. We get out to investigate. I think to myself this must be the Viking Longhouse our Lord of the Rings loving German friend must've slept in. I climb up and lower myself into the longboat complete with a couple of oars I can't help myself. I shout out I'M SPARTACUS. Not many folk will get to do that in a Longboat. (for younger readers look up Spartacus—Kirk Douglas). I reluctantly climb back out.

Then into the Longhouse. It's got a roof, long wooden seats, wooden log tables and a fire pit in the middle. I can easily picture a bunch of Vikings sitting round the fire, drinking mead (or Carlsberg) and knocking the stuffing out of each other.

We camp outside for the night and the next morning we decide that the Longhouse is the perfect place to make breakfast and dinner as a one off. After dinner, I put Margaret over my shoulder burst open the door and head for the van. At this point, she jumps down puts me over her shoulder and drags me to the Longhouse to do the dishes.

We sit and admire the view since Unst is so small there is a coastline almost everywhere you look. We notice a sign with an arrow pointing to an inlet and it says Otters this Way. Maybe we will get to see one after all.

Sitting here it feels to us that we are doing the right thing, travelling and camping, enjoying the unexpected. Tonight is a very starry night and the silhouette of the Viking Ship and it's long neck looks very impressive. Once again I think of all these old Viking films. I hope Margaret has got some Viking blood in her. I must admit she does look Scandinavian and she loves a fire to the extent I worry sometimes, especially when she starts dancing and screaming round it. I think she just might have the way she enjoys stuff like this. She is definitely Made fae Girders, especially being brought up on a farm by hard working parents. She is the eldest and also has two brothers. The whole family had to muck in from a young age, so the outdoors doesn't bother her. I'm the same. I am definitely a Nomad.

Gimme Shelter

MAGGIE:

Pete made the Prawn stir fry meal and it tastes great. We have it along with red cabbage mmmmmmm. We had a great sleep. We noticed that the boat next to our campervan was named Tinkerbell and it does feel as if she has sprinkled some of her fairy dust on us. We feel so lucky to be here.

Now, it's time to go and explore Haroldswick and the surrounding area. Firstly we head to the Boat Museum to find out opening times as it was due to close for the Winter. Once again Pete the Rock found a rock. It's called Hopeman Rock. He just keeps finding these mojos or totems wherever he goes. We have been getting a lot of cosmic signs. These signs are there for every single one of us to help guide us along our way. We have been given free will, but help is always there. All we have to do is ask then look for the signs.

We can't believe our eyes as we drive up through the village. There is a house, on the wall there is a handmade piece of art. It's a face with an arm and a finger coming up to its lips in the hush signal. It's so unusual looking and it's obviously homemade. Pete just found out a couple days ago that his Granny's maiden name was Hush. Weird. We take a photo and take it as a sign.

We went into Victoria's Tearoom next to the Boat Museum, had a cappuccino each and shared a piece of tiffin. We then drive slowly along the road looking for otters. After a wee, while Pete spots one running up the beach. I missed it, but it's a good omen. There are lots of seals basking on the rocks and we sit and watch them for a while.

The winds are picking up and there is a storm coming according to the weather forecast. We drive all the way up the road to the Hermaness Nature Reserve which, by good luck, has a car park sheltered from the winds. We park next to a building which should give us shelter.

PETE:

It feels so quiet, peaceful and remote here. We decide to follow the arrow that points to otters and as we have one last look at the big Longboat a car appears and as it passes by the driver smiles and waves to us. They are so friendly and welcoming up here.

We pass a few houses and even a small tea room which is open and have a couple of coffees. We drive on nice and slow on the lookout for otters and I see one run up the rocks in the distance. As we drive further up we come across a big complex with loads of buildings and it looks all shut up. I notice a big red T on the wall. I know that sign anywhere—it's a Tenants pub, sadly it too is shut. We find out that is an old RAF base which is now a holiday complex. Most of the season it is open for holiday makers and a constant flow of twitchers the bird watchers.

We drive on up the road until we come to a car park which stops at the sea and the Hermaness Nature Reserve and it seems that all this land is owned by Scottish Natural Heritage. We can go no further. The next stop is Norway so we are at the very top of Scotland. We decide to stay the night and hope to see the Aurora Borealis.

It certainly is a good sheltered place. We know this because it was very windy and stormy last night and our van hardly moved as we had parked next to a building which gave extra shelter. We see what looks like a small storehouse right at the very edge of the sea with a concrete ramp. I think it is an old boat ramp. The shed next to the van is big enough to hold a boat.

There is a Heritage Information Centre right above us with flats attached to them. We walk up the hill to the Centre and it is closed. There is a campervan sitting up next to the flats so we wonder is it an old abandoned one or is there actually someone living here in the last place in Scotland?

Rest and Be Thankful

MAGGIE:

The forecast proved right it was very stormy, but we felt safe in this hollow next to the building. It looks like a shed for a RNLI boat perhaps as there is what looks like a deserted wee storehouse across from us, right at the edge of the sea.

The wind has calmed down to a breeze. We got talking to some ladies and we asked about Hermaness. Seemingly, the Centre has been closed for a couple of years, but there are a couple of flats occupied, a holiday cottage, storehouse and boat house, they think both belonged to the Lighthouse keepers or RNLI.

The Boat Museum had interesting information about the Herring Girls who followed the herring shoals up round the Shetlands and right down to Yarmouth doing all the gutting of the large catches. I was hoping to sample some herring, but it is the wrong season. My Granny made great potted herring which has kind of went out of fashion now on the mainland. I did have it at a Scandinavian cafe last year in Tenerife. It was delicious and brought back memories of my Granny. Food can do that, as well as songs. Memories like that are precious.

We go back up to Hermaness to park for the night. We meet a friendly, wiry, fox terrier. He belongs to a couple who are renting the quaint little holiday cottage nearby for a couple of weeks. Like us, they are hoping to see the Northern Lights. The dog is called Speedy Gonzales. The very nice, good natured lady owner comes from Columbia, South America. Can't resist saying Arriba arriba…andele andele…

PETE:

The weather is so unpredictable here. After our exploits at Levenwick and the severe gales, we decide that parking the van near the boat shed in sheltered Hermaness is our best option on Unst. There is a shop only about a mile and half away which sells everything we will need and it has a public toilet. It is so quiet and peaceful here so we get some provisions and chill.

After giving the van a clean, I walk down to the edge to have a look at the small building there. When I try the lock, it opens and obviously has not seen daylight for a long time. There is a table with two chairs on top of it and a small sofa, all in need of a clean. Then I see a cabinet on the wall and it says 'Stress Relief' I open it and instead of bandages and medical supplies it's got four bottles—two dark rum, one Bacardi and one red wine. All are opened but it evidently has been a while ago going by the cobwebs. There is a photo of a Selkie on the wall and some fishing stuff. Again cobwebs. We guess it was an old sea cabin for fishermen or sailors.

To my surprise the light works and there is a window which looks out to sea. I shout Margaret over and we sit on the two chairs at the table and look out onto the sea. At this very remote place, at the tip of Scotland, we both decide telepathically that after a clean-up, we have found the ideal place to sit and do what we hoped to do… we are actually going to write our book on our travels to Australia, New Zealand and back to Orkney and Shetland called There and Back Again.

We have our own cooker and wee gas heater and the cabin is winter proof. We will write in the wee cabin and sleep in the van. Perfect. We have to give this place a name Margaret says. It's a God send.

It's quiet and remote and surrounded by nature. I agree and as I look around I see the cupboard with Stress Relief written on it and I think well that's true and the word rest jumps out at me. That's it 'Rest'. Hey, we might even see orcas!

MAGGIE:

Pete has shouted me over to the wee storehouse. The door wasn't locked and it is perfect for writing. Pete says he is calling it Rest and I said Rest and be Thankful. We are so excited to find such a perfect wee place to sit and write our book and not bother anyone. Thank you, thank you, thank you.

We spent the day writing finding inspiration looking out the window across the wild sea. Despite the stormy weather, we slept very well in our van. When we woke up, the storm has left as quickly as it came. We make breakfast in the wee howf.

There is a big picture of a Lighthouse. Pete got the Hope Angel Card this morning and it so happens that one of his favourite songs is called Hope by Klaatu from the album Hope and it's about a lighthouse keeper or how we are all lighthouse keepers. Seriously, you cannot make this up.

PETE:

We decide to have an early night. We go to our van with our Icelandic wool blankets, comfy rock n roll bed and feel pleased with ourselves. We can't wait to start writing. We have the table and chairs looking over the sea. I've designed an indoor lock to stop any high winds blowing the door open. Margaret has even put some photos up for inspiration.

We are going to get up early, but a noise like a car engine wakes me up and there's voices. It sounds like two men. I get ready and go out to see. They are actually here to do a spot of fishing and hopefully catch their breakfast. I start talking to them. They sort of say hello and get on with their rods and lines. Straight away one of them reels in a small silvery looking fish. A sardine I think. "You will need a few of them by the looks of it," I say.

"That," he says, "is my Mepp."

I decide to leave them to it and for some reason he tells me that a mole can tunnel sixty five feet in a day.

MAGGIE:

As we head to the shop which is called The Final Checkout we have to stop for a herd of miniature Shetland Ponies. They are on the road wandering around wherever they want. I love their attitude and I love their wild hair. They remind me of Pete: small, hairy and adorable.

We stop at the famous bus shelter we were told about. It has a Science theme in honour of Galileo the father of science. There are lots of techie things, I guess representing sciency things in the shelter. We also come across a John Peel memorial as well.

Baltasound has only about ten houses and also a cafe/shop, a hairdresser, an unlikely looking pub, an hotel, a working harbour and a Post Office. We don't need much more than that. There is also leisure centre nearby where we can get a shower.

On the way back, I saw an otter running up the same spot as Pete saw one the first day we arrived on Unst. We sat for ages watching. It feels like a privilege. Then we went back to our wee howf Rest and started writing our book. This is the life. The kettle is on the stove whistling and we are whistling too.

PETE:

As we drive to do our shopping I remember that this is 8th October which would have been my Nana Bicker's birthday. She would be one hundred and two if she were here. We drive past the longboat and there is nobody there. I keep forgetting it's Winter because the weather on the whole has been great apart from the odd storm at night.

We have to stop as right in the middle of the road, caring not a jot, are some wee Shetland ponies. They are so tiny they are only up to my waist and I'm Hobbit size. They are the cutest things ever. I think they must have been from the time of the Picts as they were small folk and these ponies would have been ideal. Maybe I'm a Pict descendant. I'll settle for that. There are two types of Shetland Pony—the standard and believe it or not the miniature. The standard is forty two inches high and the miniature is thirty two inches high.

We reach the shop which is a small supermarket/cafe. The place is full of cars and we find out that there is a garage attached. As Margaret goes into the shop I notice to the left a small, grey, square building not much bigger than a public toilet. Closer investigation reveals it's a pub and it opens whenever they feel like it. Today it seems they don't feel like it.

I head to the shop and find Margaret has bought milk and a few things. If we want sausages, bacon or any butcher meat or fruit we have to wait until 3 o'clock when the ferry brings more supplies and the van has delivered it. We decide to go exploring until then.

As we drive along we come across something instantly recognisable, even if I have never seen it before. No, not an orca. It's the famous Bobby's Bus Shelter. It's done out like a living room complete with technology such as microwave, a phone. Also has curtains, chairs and a visitor's book which I sign and date. Looking at the names and places there is a Bill from Australia, a Tom from Canada, another Bill from Wales and even a Paddington from Peru?

We sit seal watching for a while. They seem larger than normal. Maybe they are just closer. We collect some meat from the shop and head back to our 'pad'. We take it slowly and stop at the Haroldswick beach and Margaret sees an otter running up the beach. This time I didn't see it.

We get back and start writing. I feel like a real author sitting here in this remote building with a small light on, facing the sea and the stars. We write well into the night. The sky looks like a bit of a lime colour to me so I am hoping that it's the Northern Lights.

MAGGIE:

We slept very long, so we must have needed it. We drove over and up to the top of the hill so we can see across to our wee howf. Amazing views. We feel very high up as we look over and see Muckle Flugga Lighthouse and Hermaness. Later, we had mince, tatties and carrots for dinner and then we wrote.

My goodness we did it again. We slept till 11.30 and then we snuggled in until 14.25. We must need it. Lots of healing going on. Either that or we are in hibernation mode. We had the latest breakfast I can recall ever having. Then we did some writing. It felt great and it felt fun. So looking forward to writing every day with Pete.

I was reading the Pamela Stephenson book and thinking about Billy Connolly and the joy he has brought to this world. His show The Sex life of Bandages is showing in cinemas today and tonight around most of the UK (we have to wait until the 19th up here) I believe in the collective consciousness and the power of that collective consciousness. I also believe in miracles. I believe in the power of prayer, especially after our experience in Kawakawa. I wonder if I say a prayer for Billy at the point of lots of people watching the film, everyone will be laughing and feeling the love at his stories. Would this help give a healing through the collective consciousness? Wouldn't that be marvellous. He is a national treasure who deserves good health. He has been through such a lot in his life and yet he brings us joy and the ability to laugh at ourselves. I wish you well and all the love in the Universe, Billy Connolly.

PETE:

I study a map we have with us. I know there is a Lighthouse near us called Muckle Flugga. I also discover that there is a tall Stack (that's a tall rugged rock standing straight up out of the water) The Stack is near the area of Tonga! Well, my mum went to Tonga, but it was the one in the South Pacific, and she wrote a journal of her travels. Without even knowing that there is a Tonga in Unst I brought the Journal with me, so I am going to read it here, then go a walk tomorrow to Tonga. I never saw that coming. Another reason I am going is that the map also says that the small streams have gold in them.

We didn't get to Tonga today as it was far too wet and windy. To get there you have to walk through moors and heather. Get lost up here and you've had it because no one will know you have been here. We decide to drive up to the highest cliff which we can see across from us.

We drive up the windy narrow road. When we get there, there are lots of giant radar discs and empty abandoned buildings. We are very high up and you can see over the sea for miles and miles and behind us looking at the scenery it's no wonder they say Scotland is like New Zealand. It is beautiful.

Margaret points out a large bird. It could be a raven or a buzzard I expertly say. No says Margaret it's a Bonxie as they call them up here. A Great Skua. We head back down the steep drive and go to what's becoming a regular little spot for us along quite near the Viking longboat to look out for otters and seals and our nightly dusk time watch. Tonight we are treating ourselves to a Black Velvet to help celebrate the launch of our careers as authors, travellers and wildlife spotters of the World. Still no orcas...

MAGGIE:

Going to Uyeasound to the Hostel today to see if it is open and hopefully do a big laundry. Found a good shop on the way down the seven mile journey to the bottom of the Island. It sells a good selection and has its own bakery. We get a Stornoway Black Pudding and I shout "Ecky Thump." I always wanted to do that. We also get some Sassermaet which is a Shetland variation of mainland square slice sausage, except tastier. Having Stovies today and a fry up tomorrow. It's the winter and we need sustenance. That's my excuse and I am sticking to it.

The Hostel is much bigger than we expected. There are still a lot of twitchers about so it is open. Got my wee binoculars out and they look pathetic next to the huge telescopic lenses of the twitchers. They are a nice bunch of folk though. Haven't got to their level of twitching but getting there. I love birds. We had a nice day and sat in the Conservatory. The twitchers all came back to roost after twilight. Pete was talking to them in the kitchen. A lot of them are going home in the morning. There has been a lot of rain today. I just kept on writing.

Next morning was nicer and we find that the fairies have been. Someone has put a full bottle of wine and a three quarters bottle in our box of provisions in the kitchen. We got talking to Mungo a very decent, friendly chap who is born and bred in Unst. His granny was a herring girl and his grandparents were the last folk to leave their remote croft (out of fifteen crofts) up past Haroldswick. He was brought up by his grandparents on the croft and it was a hard way of life. He was also involved with the Muckle Flugga Lighthouse. Well he said he had to 'love us and leave us'. It was a pleasure meeting you and hearing your stories.

Lewis the Lighthouse Keeper

PETE:

The next couple of days, we enjoy writing our book and taking a break now and then to explore this small island even more. I read a pamphlet that says there's gold been found in the running streams of Unst; I'll definitely keep mind of that.

One evening, we were just about to stop writing and get ready to go to the van when we both jumped out of our skin. There was a knock on the door. I mean, we're in an old store shed at the very last piece of land in Scotland with Norway as our neighbour, the last thing we expect is a knock on the door. I answer it wishing I was a lot taller than five foot three inches. There is a guy standing with a light on his head and a beard. Sorry, he said, I thought you were somebody else. He went on to say that he stayed in the flat in the Old Visitor Centre and had seen the light on in the old storehouse these past few nights. He thought we were an old friend and he was coming to say hello and with that he left.

So, we say, we're not alone away up here. What will we do as we don't want to leave the peace and quiet just yet. This is the ideal writing place for us and we are not bothering anyone. We decide we will talk to him in the morning and explain that we are writing a book on our travels in a campervan There and back Again...Australia—New Zealand back to Orkney and Shetland.

Next morning after breakfast and the short walk up to the Old Visitor Centre, we knock on the door we think is occupied. He opens the door and we say hello and explain ourselves. He tells us his name is Lewis. He tells us he used to be one of the Lighthouse Keepers on Muckle Flugga Lighthouse which can be seen. What an absolutely awesome name Muckle Flugga.

He must like us, to our surprise he offers us the use of his shower when we feel like it and his toilet. He will keep the door open for us. Let me make you a home-made curry sometime. We leave feeling very happy to have met him.

We have decided to visit a Youth Hostel at the bottom of the island to see if it is open. To our surprise it is and it's a great place with great facilities as we want to do our laundry. Also to our surprise there is a bunch of folk staying. They are the famous twitchers and they told me all about the birds they'd seen.

After breakfast, we were washing our dishes when the cleaner came in. We started talking and he asked if we were staying long. We told him we were travelling but are actually staying in our van at the old Nature reserve. Well, he said you'll know Lewis. I worked with him on Muckle Flugga Lighthouse and also at the Salmon Farm and also at the Airport. He tells us that his cat regularly walks two and half miles from his house to Lewis's house. I can't wait to tell Lewis we have been speaking to his friend Mungo.

MAGGIE:

Before we left the Youth Hostel, one of the twitchers came into the kitchen. They were just heading off and he gave us some porridge and some onions. They had been looking for a Grey Shrike at the early hours this morning. He said he is going home to 'auld claes and purritch noo', but that he has been giving his liver a holiday as he goes tea total every time he goes twitching.

On the way back up the road, we decide to go for a walk on Sandwick Beach. It is a beautiful beach with white sands and the tide is out. There is an old Viking Settlement here with a longhouse right at the beach. They reckon the Picts were here before the Vikings did away with them.

Then we can't believe our eyes there is a big male otter heading down the beach. He has quite a long journey because the tide is out and the wind is in our favour as we watch him slide into the sea. I think what goes in must come out and so we hang around for a while, but there is no sign of him.

We decide to walk round to an old Norse Church ruin and which is still used to this day as a cemetery. Bodies are brought by tractor and the mourners have to walk along the beach. Nice place to be laid to rest. I like the thought of the walk along the beach.

We decide to head back slowly. I see a seal right close to the shallows as if trying to catch my attention. I had just mentioned to Pete that something had spooked some ducks on the water. They rose up and flew around in a circle to land back where they took off. Then there he is! Coming out of the water, with a large something in its mouth, is the otter. We saw a wing flap as he dragged his catch up the beach. I had my binoculars and what a sight in broad daylight. The otter had actually caught what we later found out was a Skoter (I looked up the bird book).

PETE:

As we left the Youth Hostel we saw a signpost which said Sandwick beach and Viking Longhouse and ruins. We felt we just had to go. As we drive through the village of Uyeasound we pass a primary school and on the walls the kids had painted rainbows, dream catchers, Vikings, poems, all of which looked great.

Once parked we get out the van and start walking towards the clear and golden beach. It looks fab and it's on the way to the old church ruins.

We pass what looks like an old croft cottage which is in great condition. The door is open but there is nobody there. I give a shout 'Hello' but nobody there. We are quite remote and I shout again. It's like the goldilocks scenario again with plates and cutlery sitting. I decide to go in. There is a bottle of wine on the table with an envelope lying open with money sticking out of it. It's not sealed but I am guessing this is a holiday cottage. There is a note on the table saying, "Thanks Cottage for a wonderful time." The date on the note was a week ago so it has been lying around all week like this. It just goes to show how trusting the folk are up here. It is a nice feeling harking back to the old days. I still decide to close the door when I leave. I tell Margaret all about it and we are both sure that the right people will get it OK. It's a great feeling.

We move down towards the beach and I get to see an otter. It's a great view and we get to see it running down the beach from the stream to the sea. Otters like freshwater even although they fish in the sea.

We move on to the old church and cemetery. You can feel the energy of the old church. There seems to be the same family names on the headstones. As long as it's not mine. I am not ready yet.

On the way back, Margaret points out that something has spooked a small flock of ducks sitting on the sea as they all rise up. Margaret lets out a shout, "Look over on the shore over there." We can see the otter coming out the water with a duck in its mouth. A duck!. It trailed it up the sand towards the stream. The duck seemed huge in its mouth and it's wing was flapping.

I decided to follow its tracks. I stayed downwind and followed the track till it disappeared. I saw the old bones lying around so I knew I was at the right place. I moved very slow and then there it was, the otter was eating it's lunch right in front of me. Suddenly it saw me and ran off. I stood for a moment. I had a feeling that this must be very rare to see, even to an experienced wildlife camera man. I moved on. I think the otter would return later when it knew we had gone.

MAGGIE:

I was taking photos of the tracks on the beach. I never thought for a moment we would see him close up. They are notoriously shy. Pete was trying to get my attention. He had followed the otter track and had come across the otter actually eating his catch. It sensed him and made a sharp exit. What a sight that must have been close up.

The tracks reminded me of our turtle expeditions on Heron Island. We 'WhatsApped' the lassie from Scottish National Heritage later in the week to tell her what we had seen. We certainly learned a lot. I had no idea that otters eat ducks or birds.

As we head back along the road we pass the Galley Hall of Uyeasound. I happened to notice it was open and did an emergency stop. We asked if we could peek in and were invited by the next Jarl of Uyeasound Up Helly Aa to come in and he would show us around. He was sporting a big bushy beard which I think must be compulsory to be Jarl. It was very interesting to see all the different shields, all handmade. Each one was displayed and also photos of each year's Jarl. This guy's dad was Jarl in 1993 and his grandpa in 1972 so it is a family tradition. I would love to stay and see Up Helly Aa.

We got to speak to Lewis who had knocked on the 'Rest' door a couple of nights ago. Turns out he was one of the Lighthouse Keepers on Muckle Flugga during the 1970s. He has lived here most of his adult life as well, apart from when he was married. He must like his own company.

PETE:

We got in the van and drove off chattering excitedly about what we had just seen when suddenly, the brakes got slammed on. My head nearly went through the windscreen. Then, just as quickly, the van reversed as well as my head. Thank God for seatbelts is all I can say.

Margaret had spotted 'The Galley' that's where the locals build the Viking boat to be burned at Up Helly Aa. They also make Viking shields and leather or sheepskin waistcoats. We looked inside and the guy, who looked like a Viking, showed us all around. We saw the boat, the shields, the clothes and photos of past Jarls up to the present. He is going to be the next Jarl. A Jarl is the guy in charge of the Viking march and burning of the boat. They all have their own distinct design they make up. He shows me his axe and then he offers to take my photo holding the axe and shield and putting on my best Jarl face next to his Viking boat. I thank him and we move on.

On the way back, we thought about what we had seen on the beach. We had heard a rumour that there were island sheep which ate seaweed and tasted amazing. Seemingly it's much different from ordinary lamb and is an island delicacy. Well, it's not a rumour, it's true. We saw some today. They were wandering up and down the beach munching seaweed for ages and did I mention we saw an otter bring a wild duck out of the sea and I saw it eating it. All in all it was a day in the wild to remember and one for my Nature Journal.

MAGGIE:

When we wake up next morning at Hermaness, it is a lovely day. Perfect for walking over the Nature Reserve to see if we can find Tonga. It would be good to get a view to Muckle Flugga too. We take a wee picnic of peppered mackerel fillets.

The top of the moors is nearly as high as the old Radar Station opposite. Pete sees lots of otter spraint around a bog which has wee pools of water around it. There is definitely evidence that otters have been feeding here. We are constantly learning things about otters up here in the middle of nowhere. The river is quite a distance away and the sea is a sheer drop from the cliffs and yet they are up here on the moors. It is so fitting that Pete won a Nature Journal in the raffle.

Going by the map we reach the Tonga area, the spectacular Tonga cliffs and Tonga Stack. It seems to be a bird nesting sanctuary, but then the whole area is. This is an important moment for Pete because of his mum's love for the other South Pacific Tonga and he tries to phone his stepdad. We walk back over the moor and have our picnic overlooking a loch. Then to write...

Weet Breeks

PETE

Last night when we drove back up in the dark towards Hermaness. There's a figure in a white robe and hood overhead standing at the side of the road. This gave me the creeps the first time I saw her. I thought it was a ghost. Later on we found out all is well. It is a statue, albeit a spooky one. Seemingly there is a ghost called The White Wife of Watlee and the statue, erected in 2019, is a memorial to her. The story goes apparently she lost her son and is patiently waiting on his return. There are a few folk on the island swear as they are driving along in their vehicle, out of the corner of their eye they see her sitting in the passenger seat AND she only sits next to single men probably hoping that one of them is her son.

This time I'm not scared. I know it is just a statue. In fact, I stop the van and go for a real close up. She stands with hood up and head bowed in silence. I can't resist a look under the hood. YIKES I wasn't expecting that. The face looks as if the skin is half eaten and rotten, eye balls staring and long straggly hair. I'll probably dream about her tonight. I could kick myself for looking…

Last night to take my mind off the white wife, I thought about the shopkeeper at the garage near Mid Yell before the ferry to Unst.

"Where are you going?" she says.

"Unst," I say.

"Well, why don't you park your van here for the night because tomorrow the diesel goes down five pence. You'll save a fortune."

We get a good night's sleep without any ghostly visitations. As we are having our breakfast, I hear young children laughing and shouting. There is a family down at the small beach splashing and having a good time. Then I clearly hear the mum saying quite loudly,

"Right up yi git n weel tak aff yur weet breeks." It's a great accent. I could listen to it all day. The children were having a ball just throwing stones in the water and paddling.

We met Lewis again today. He invites us to lunch soon and we find out that he likes alcoholic ginger beer. A man after my own heart. I love it too so that's that settled. Ginger beer and a homemade curry coming up soon.

I tell him we drove up to the old Radar Station and he tells us that it was built during the War to withstand 150 mile an hour gales. Two of them got blown down during a storm. Shows you how windy it gets up here. It was bad enough when the catch blew off our roof case. He then goes on to tell us that him and his mates from the Lighthouse days used to sit around the fire telling stories.

One of them was 'who has been out in the windiest days'. He tells us to be careful if we are walking to Tonga as one friend told him he was walking on the moors up there when a gale got up and he says that the gale was so fierce that it blew his mouth open and he couldn't get it shut again till the gale subsided hours later. I like Lewis.

We start off walking towards Tonga. I am quite excited because, as I mentioned before, my mum went to Tonga, but her Tonga was in the South Pacific. Man what a dream come true that must have been. I am going to read her journal of it. I brought it with me. In fact, some of my mum's ashes are scattered in Tonga. They are also scattered in Cambridge and Prestwick. My mum (bless her wee cotton socks) was born in Prestwick, moved to Cambridge and whilst there met my step dad.

When the Queen of Tonga came to the UK for the Queen's Coronation in 1953, seemingly the whole country fell in love with her. My mum wrote to her and got a letter back from her Lady in Waiting. She corresponded with the Lady in Waiting for many years till eventually she was invited by the King and Queen of Tonga to visit them over there. My mum saved for three years and she and my stepdad went over on a memorable journey where they were welcomed and treated like Royalty.

They still kept in touch and at my mum's funeral service there were three Tongan women there showing their condolences dressed in full island regalia.

When I brought the journal with me, I had no idea there would be a Tonga way up here at the opposite end of the World to the South Pacific Tonga. I was very surprised to see a Cambridge in New Zealand but to see a Tonga in Shetland Islands, well, awesome!. Small world indeed. We reach the cliffs and big Rock Stack that's Tonga and wonder how it got its name. I know what I'll do. I'll ask Lewis.

We stop and soak in the views at Tonga and when I look in my rucksack I have forgotten to bring the Journal. It's a bit too windy so I'll read it back at the howf. Despite the weather up here, we were told not to wear waterproofs as if you fall you can slide and keep sliding like a sledge over the cliff. We have seen lots of otter spraint and tracks. We learned that's what otter's poop is called— spraint. We learnt that at Mull. We end up a wee bit lost it's really remote up here. I come up with a good idea we will follow one of the streams as it will lead down to the loch. There is a fence the shape of the cliff to stop going to near the edge. When I look over, I see thousands of seabirds.

We move on as it's a bit too windy for my liking. We follow the burn through the heather moor. The burn is only about two or three feet wide and you can see stones through the crystal clear water. We feel a real feeling of adventure in the open air and nature as we hear the water trickling.

We see a fairy like dell with ferns, moss and rocks. In the water, there is something glittery catching my eye. I get to shout the immortal words There's Gold in them their hills! Man I have just seen gold with lots and lots of little bits in the rocks. When I pick them up out of the water they look brown, put them back in and they turn gold again. I don't care. This is great. I wish I had my Australian bush hat on.

We walk further down and see more. Some breaks in my hand and some looks like gold. I take lots of pictures and put them in a bag for luck. Well, you never know. We take one last look before we head down to the loch.

MAGGIE:

Pete had a lot of fun looking for gold in the burn. It certainly is a geologist's dream up here.

In between writing, I read Pamela Stephenson's book Treasure Islands which is very enjoyable. A lot of trickery went on in history and many islanders were duped into the slave trade often leaving only the elderly and the young on the islands.

We are going back to the big city or as big as it gets in Shetland. We are off to Lerwick. We got talking to Lewis for ages. He has some real good and funny stories to tell so we ask him if it is OK to put him in our book. His stories about Peerie (Shetland word for little or wee) Willie Skollar are really funny. Peerie Willie refused to let Lewis tape him, so all the stories are passed on in the oral tradition. So our book is preserving Peerie Willie stories for posterity. We are looking forward the meal and a blether next week when we come back up to continue writing. Having a couple of black velvets tonight.

Cat Brew

PETE:

I fall asleep dreaming of what I am going to do with my new found wealth. After breakfast, I feel lucky because of my gold rush yesterday. I mention to Margaret I fancy going to the shop, getting a paper and putting a wee flutter on the horses. Eh says Margaret surely they don't race those cute wee ponies. Honestly, I do worry about her sometimes, but then I see from her face she is taking the piss.

When we get to the shop, we discover that newspapers don't reach the island until the tea time ferry arrives, so no horses today then. Then, as if by magic, a big herd of tiny ponies come galloping across the field opposite. We go up to the fence and they really are tiny, just up to our waists. We get to clap and stroke them. They've got lovely long manes and nice colours. We take some photos, a video and move on.

We stop in at the Viking Longhouse. There's still nobody else there and so we go in and sit on the logs. Margaret says I look like a Viking so why don't I etch one of her blank rose quartz rune stones and she had lost one and needs it replaced. Great idea I say and sit on the log feeling right 'Vikingy'.

A Stanley knife is the only tool I have and I realise it is not the right one so I will need to do it some other time. As we pass the otter sign we can't help thinking how lucky we have been to see an otter go into the sea, catch it's dinner, bring it back out and see it eating, even if it was a duck.

We reach our little writing shed and meet Lewis again who reminds us that we are invited to lunch and a ginger beer next week. Ginger that's a brownie point. We talk for quite a while and I tell him that we met Mungo who told us that his cat walks miles coming over to visit. Lewis says yeah and it brings fleas with him. I've got him a collar and he is not allowed in the house any more. I also mentioned that Mungo worked with him on the Salmon Farm and he said yeah I can't stand salmon because of that job. That'll be the fish supper hit on the head then.

Before the Muckle Flugga Lighthouse got automated, there were three men on it. Lewis, Mungo and some other guy. On their day off, they would visit each other along with other locals such as Peerie Willie for some potent home brew. He told us that on one particular night they filled up the basin and sat around it filling their mugs. Unexpectedly, the cat fell into the basin. The story goes that his friend grabbed the cat by the throat and slowly ran his other hand down it's body to the end of its tail saving a few glasses. He then said don't be surprised gents if your next glass has got a few hairs in it. Lewis has got a few stories. We're looking forward to our dinner with him. He's also a keen photographer and he says he has some real good photos taken up here. One time he was looking at the Aurora Borealis when a meteor shower flew across. Now that's what I call cosmic.

We're going back to Lerwick tomorrow as it's time for Billy Connolly show. We say to Lewis we'll let him know when we are back and he says enjoy your night at the Pictures...

To get to Lerwick we have to drive across Unst to the ferry, drive across Yell to the ferry, drive off the ferry to drive an hour back down to Lerwick to go to the pictures. Drive, Ferry, Drive, Ferry, Drive. I am sure you will agree Billy Connolly is worth it...still no orcas.

MAGGIE:

We are going to see the Billy Connolly show the day after tomorrow. We get the two ferries and then an hour drive on the Mainland down to Lerwick. We can't find a decent chippie in Lerwick. We thought we would be spoilt for choice. It's still nice to be back. We headed to the Broch and then back to our favourite wee harbour. Lovely clear sky.

Next day we head to the Library. I find a book by the Scottish poet Jackie Kay called Red Dust Road. Feel it was meant to be as Pete got an arts newspaper and her name is the first thing I see mentioned along with a photo of her. SIGN. It's about looking for her real parents as she was adopted. Quite topical with what has happened to Pete and what he has just found out.

We went to the Family History building to see if we could find the Sunday Post article about his Granny Hush getting her arm amputated without anaesthetic. We can't find it. Pete remembers seeing a cut out of the article years ago which was on the Francis Gay page, but it has been lost somewhere along the line. We will need to contact them I think.

We find a family history website with a free fourteen day trial and sign up. When we research Pete's granny, it's a bit of a mystery about his dad. We do find her in the 1901 and 1911 census. In 1901, she is living with her gran who was head of the house then. We think her mother age twenty four maybe down as her sister? Strange. By 1911, Georgina is staying with her 'sister' and eight siblings and there is a male lodger. Pete found an article about his granddad Andrew Clement Burleigh who died by gas poisoning down the pit at Glenburn Colliery. He managed to survive the horrors of War, but sadly still got gassed.

We end up at the Mareel cafe bar to write our diaries and read our books and then it's time for Billy Connolly and he does not disappoint us. A week later we are still laughing about his Bonnie Prince Charlie story about Dumfries. He looks great on the big screen as he sits sipping a cup of tea. What a gift he has and he has given us.

My Favourite Librarian

PETE:

We time it so we reach Lerwick at night and camp at the Harbour. Shetland parking is second to none. As long as you are tidy and respect the Island there is enough parking for everyone. We don't see Billy until the day after tomorrow so we've got all day tomorrow to do something.

We decide we will do some ancestry/family history, visit the lovely library and then a nightcap and some writing at the Mareel cafe bar at the Arts Centre. Firstly, we go to check out our Ancestry to see if we are Vikings.

Well whilst we are checking the records something strange happened I came across an article in an old paper relating to my granddad's death. He was found gassed down the pit. He was only thirty six. I remember my dad telling us that he had wanted to be a journalist, but when his dad was killed down the pit it left him, as the eldest, the breadwinner for his mum and the rest of the family. So he had to work, first with the railway and then the pit. My gran never went with anyone else again and she lived until she was eighty six. I think my twin brother gets his writing bug from my dad as he is quite a poet with his prizes and such. Me? I take after mum. I hate being indoors, love camping and travelling. I am so glad and lucky that Margaret is the same.

We revisit the library. Every time we go in it's a joy. It really is an impressive building and the staff are wonderful. As I walked round a young member of staff asked me if I liked the library and did I use it often? I spoke to her for a while and she asked whether I would mind getting filmed and recorded for feedback to Headquarters. So we sat down in front of the camera in the staff room, in front of some books and I told my story in the hope that it would help.

I told them that I use the library a lot and back home I was always in it. I was so grateful for the free use of computers and coming from a small village the library was everything. I told them I did an online course and sat exams in the library. I couldn't afford a computer and the library was really a way for me to raise my life and sit my exams. I told them that the staff were very helpful and in particular one wonderful Librarian

1. I passed all my exams.
2. A few years down the line I married the Librarian.

Perfect says the interviewer if that doesn't save a library then I don't know what will.

Before we go to the Broch, we have another check of my family history. This time we find out that my granny was not Scottish but born over the border in Berwick on Tweed. We find out that according to the 1911 census that she was a servant. Still no sign of a Birth Certificate for my dad. It's as if he didn't exist. What was it the girl said…when researching family history…BE AFRAID, BE VERY AFRAID.

We watch Billy Connolly at the Mareel Picture House and it's very, very funny. I highly recommend the DVD. The pan pipe sketch and Bonnie Prince Charlie shoes sketch are hilarious. It is taken on his last Australian Tour The Sex Life of Bandages. Fucking brilliant.

MAGGIE:

The weather is being kind to us and we are visiting the Broch as much as we can. The energy is great. We spot an otter swimming in the sea near the Supermarket. Another couple with a huge camera have seen it too. It comes out onto the rocks and makes its way up the shore towards us. We get a real close up view and he even looks us in the eye. Unfortunately the woman stepped on the rock above him and he ran off, but we feel very lucky to see another otter, in the wild, so close up.

As we make some burgers in a sheltered spot near Knab a taxi driver stops to chat to us. He tells us he saw the converted taxi camper van and that he loves wild camping. Him and his wife had been travelling Europe for months and spent three months in Morocco. He is only just back on Shetland but looking forward to more travelling. Two years ago in January he lost his dad, then his mum died the next day and he reckons his mum died of a broken heart. Very sad. It does teach you to live each day as a precious gift. He says Shetland is a camping friendly place and we totally agree.

PETE:

We decide we want to go back to the Broch to see if our phones get zapped. We go a walk along the loch next to the Broch and I spot what looks like a Native American canoe which could sit about eight braves or squaws. Bit unusual to find an Indian canoe way up here I comment. Bloody eyesore, It was a Canadian couple who had it but they moved back to Canada and donated it to Lerwick and it's been at the same spot for the past five years. Bloody eyesore you can have it he says again. I really wish I could. Imagine having your very own Last of the Mohican's canoe?

As we walked along the rocky shoreline suddenly Margaret puts her arm out and says shhhhhhhh. At the same time, one leg is bent and an arm is doing a stop sign. Then she whispers, "Be vewy vewy quiet. I tot I taw an otter." So I adopted the same pose and, sure enough, there was a great view of an otter, quite large moving out of the sea and onto some rocks. Then moving in and out of the crevices looking for something to eat. It disappeared, but to our delight it appeared again right in front of us, giving us a great close up and even eye contact. We were so close I felt I could touch it's tail. It felt it was showing off just for us. Sadly though, still no orcas.

We sit and have a Moroccan mint tea in the cafe bar before heading to the van for bed. I notice the girl at the next table takes off her jacket, then her cardigan, she then unbuttons her blouse and pulls her arm out of her sleeve. All the barmen seemed to have stopped what they're doing and are looking her way. She then turns to her friend and shows her new tattoo off. All the barmen are not impressed and go back to pulling pints.

MAGGIE:

We leave for Unst on Wednesday because the winds in Lerwick are forecast 40 mph and in Unst 20 mph. We manage to get the 8 pm ferry to Yell and go to the Burravoe campsite for the night and do our laundry the next day.

Lewis sees we are back the next day and invites us in for coffee and very nice filter coffee it is too. We then go to the Charity Shop and get some books and DVDs. One is called Tracks and is based on a true story. In 1980, Robyn Davidson did a two thousand mile trek across the Australian desert with four camels and a dog. Highly recommended. I fancy reading her book which became a bestseller at the time of publishing.

It really is a fantastic place to write. So quiet and we feel so blessed. We are going to Lewis for a curry later and so we nip into Baltasound to get some Ginger beer to take with us. Not as good as Ginger Joe who has really spoilt us. Lewis' curry is delicious and fruity and so are his stories. When we go back, we do a bit more writing. I believe more and more that everything is divine timing and we are having the time of our lives on our travels. Freedom.

Next day we go to the Farmers market at Baltasound Hall and get a lovely plate of lentil soup and some cakes. We arrived quite late and just as we are about to leave folk come and kindly say fill these two boxes with cakes as they will only go to waste. I buy some apple crumble and a jar of beetroot chutney. Back we go to write and Pete makes crispy bacon rolls with beetroot chutney and they are delicious.

By 9 pm, our feet are freezing and it's time to go and watch the second part of The Last Samurai DVD. When I got it yesterday, I didn't realise Billy Connolly was in it.

PETE

Well, that was a treat. I don't eat curries as a rule, but that was delicious and I even have seconds. I got my first taste of coffee from a coffee bag as well. It's like a tea bag only it's a coffee bag. Genius.

After telling him about the Billy Connolly film, he tells us a few more stories. One day he was on the mainland and it was Grand National day. A local saying is "I'm not one for gossiping but wait till you hear this?…" Well, a couple he knows the wife was pregnant and due anytime. The father to be had gone off to put a bet on the Grand National, a horse called something Boy. A few hours later his wife gets a text from someone saying sorry to hear of their sad news, condolences, if you need anything we're here to help. Turns out that someone outside the bookies caught pieces of the father to be conversation. Just lost…boy. Not to worry, will try again next year…

He told us tales of his Muckle Flugga Lighthouse days and the passing whales. His days working in the fish factory and the airport. He went on to tell us stories about the White Wife Ghost and a couple of his reputable male friends claim to have seen her. Seemingly she is quite harmless. Oh well, that's good news.

I like Lewis. I know he would have loved to have done what we have been doing this past year. He liked our stories of Australia and New Zealand in the campervan. I am sure we will be back for lunch again. I gave him one of my Paua Paua shells from New Zealand. If you delicately take off the calcium, salts and other stuff that gathers on the shells on the seabed, then polish them, they are stunning. They often get used for jewellery.

After seeing the Billy Connolly film, I feel I have just got to mention this. We are going to send Billy and Pamela a copy of our book 'There and back again' especially to Billy. The reason being before we set off on our travels there was something briefly said about him the day we left. I mentioned that the pair of us have a lot in common, the swallow tattoos on our hands (I got mine way before Billy), the way we were brought up and the things that happened to us (both our mum's left us when we were youngsters). We also spoke to each other in 1973, just the two of us a great story and very typical Billy. It turned out that everywhere we went in Australia and also here in the Shetlands we end up getting books he wrote or Pamela wrote. Even on Heron Island miles from anywhere someone was reading one of his books: his autobiography. We feel he has been with us on the whole journey and would like to thank Billy and Pamela for all the good reads on our journey and hope you both enjoy reading our book.

MAGGIE

The full moon is doing its thing again. I dig deep and go with it. Most of us are walking around quite numb and in denial of our feelings. The secret is to breathe properly. My lungs are very painful as they let go years of anxiety, but I know it will be worth it. Years of living in fight or flight where we tend to shallow breathe. If only we all just slowed down and take deep breathes, life gets so much easier.

Anyone who is on mind altering drugs of any kind including prescribed medication for anxiety, are not dealing with their feelings. Once off the drugs there is no option but to be honest with yourself. I know this from experience as I had to go on prescribed medication for anxiety and depression for six years during the menopause and at the time my marriage break up. They don't call it the Change of Life for nothing. It was very difficult at the time and now I find out that I had been in avoiding dealing with emotions from childhood trauma as well. I thought good girls don't get angry or have negative feelings. Everything came to a head. I am pleased to report that I managed to come off them last year. I feel really lucky to have good health and a clear head. It was my heart that had to be mended. Living this quiet existence is a gift we should all give ourselves.

Makkin Belts

PETE:

We went to the shop and saw a poster on the noticeboard saying MAKKIN BELTS 29th Oct Baltasound Hall. I say to Margaret let's go to it and I will take my real Italian Leather belt with solid brass buckles and who knows I might learn to make or makk a belt or design or engrave some runes on my belt. I got the belt in a charity shop in Stromness Orkney for 15p. A real bargain. Have I told you I love charity shops and I am shit hot at getting a bargain? I text my two brothers and two sons to tell them about it and tell them will keep them posted.

Well. That was a bit of a red neck. We went into the Hall and there were a good number of woman sitting in a circle with two of them sitting at a table. They wave us in and tell us to take a seat. I thought we had walked in to some sort of meeting and sure enough the woman at the table says right this is the Minutes from the last Meeting discuss the past meeting and what to do this month including a game of walking netball? I look around and I don't see any sign of leather belts. I am just about to nudge Margaret to get up and go when I notice she has spotted a table in the corner full of sponge cakes and I know from experience we are going nowhere.

As the meeting finishes and the women introduce themselves they all ask what are we doing up here at this time of year. We say we are camping hoping to see the Northern lights and it's hard to tell if that is their normal look or if they think we are mad. All of a sudden the door opens and an older guy walks in. The lady at the table does the introduction and explains there will be a demonstration on how to make a Makkin Belt. There is no leaving the room now especially with the cakes in the corner. I listen eagerly.

The guy stands up and gives a little cough and says. Hello everyone. Erm, well I am going to make a Makkin Belt. So what we have here is a template, it's oval shaped. You sit the leather like so he says then you get a sharp knife and cut around the template and well, erm, that's about it really. Then he passed it round.

We find out that a Makkin Belt is used for knitting and all the women in Shetland use them when knitting to keep their knitting needles in place. We also find out that you need two pieces of leather cut into shape and then stitch them with leather, then filling them tightly with horsehair, then you stitch a strap or belt on it and wear it around your waist. Then you stick your knitting needle into it and it's used as a sort of third arm when knitting. It looks quite cool really.

MAGGIE:

We do some writing and then go for some shopping to Baltasound. We notice a poster advertising Makkin Belts. Pete fancies trying his hand at making a belt and also taking his wee Viking bag with him to see what the guy thinks. There are lots of things happening on Unst. There is a great community spirit and a generosity of spirit. We feel as if we are getting to know folk as we meet the same folk at the Charity Shop, the Farmers Market, the shop.

We are biding our time until the Community Hall opens, so I sit on the internet for an hour. Since Pete is reading Treasure Island and I am reading

Pamela Stephenson's Treasure Islands… I decide to look up Robert Newton the actor who was the best ever Long John Silver and kind of inadvertently does a brilliant Keith Moon impersonation into the bargain. I was sad to find out he was an alcoholic and died age fifty. I then look up Kit Carson, who played young Jim Hawkins in the same film and was also the voice of Peter Pan in the Disney film. I find out he died of a heroin overdose, all alone, in a squat, at the age of thirty one. So sad. Fame and fortune does not necessarily bring happiness.

Anyway, at last it is 8 o'clock and the folk are arriving for the WRI. We are made very welcome and when I offer to pay we just have to buy some raffle tickets. I notice there are four splendid looking Victoria sponges for the competition.

It's time for Makkin Belts. It is a strange looking shape he is cutting out and I am trying to figure out what kind of belt it is. It gets holes punched, leather stitching and then stuffed. I am scratching my head and just about to ask when all is revealed. It is a knitting belt and the proper name for it is a Makkin Belt.

Seemingly they are traditional and have been used for centuries. What a laugh when we explain ourselves.

PETE:

To Margaret's delight we find out that the old guy is also here to judge the cakes as it is a Women's Institute meeting we are at. They don't get passed round and Margaret is looking a bit peeved the until door opens and a trolley full of cakes, scones etc. is wheeled in.

There is a raffle and since there are only thirteen of us we buy five each hoping to win a bottle of wine perhaps. To our amazement the first ticket drawn is mine, my prize comes over and it's a jar of chutney, then Margaret wins and hers is a small bag, the woman gives it to her and says there are two balls in that bag. The women burst out laughing. I hang my head down knowing what's coming and sure enough Margaret whispers I've won a bawbag! To our embarrassment we won another five prizes including a packet of hobnobs, which I love, and a cake. By the way, the two balls in the bag I have since found out are bath bombs. It turned out a great wee night with everybody giving us a warm welcome having a laugh at our misinterpretation of Makkin Belts. We also got an invitation to tea. We think we will go because she will probably have plenty of stories to tell, because as they say up here, "It's nane o ma business but have you heard…"

MAGGIE:

The Hostess Trolley arrives loaded with goodies, one cake looks like a Loch Katrine cake, like my mum used to make. It has a kind of pastry at the bottom then fruit and then icing and it's delicious. The women are so nice and welcoming and we all have a good chat. We got talking about the Northern Lights or the Mirrie Dancers as they are called up here. One girl said when she saw them once she was lying on her back watching and it felt like they reached down to touch her. All in all it was a great night and we gave everyone a good laugh misinterpreting the Makkin Belts. We also found it very interesting.

PETE:

Last night we just missed the Mirrie Dancers. Someone told us that they were out while we were sleeping. We have an app on our phone telling us where and what the probability is of seeing them.

I took a walk up to where Lewis stays. The all-round views are spectacular. There is Lewis living here and one other person who seems to come and go. I reach Lewis's flat to find he is not in, so I sit and watch the sea for a while in the hope of seeing an orca. I hear a noise and when I walk round the corner, to my surprise, it is a coal man. I mean an actual coal man with forty bags of coal for Lewis. I wasn't expecting that way up here at the very last outpost of Scotland. I give the coal man a hand, but we have to sit them in front of the coalhouse as it is locked. Lewis always keeps it locked since although he is virtually isolated someone else keeps stealing it and anything else that burns. Lewis says as no Buddhist would steal and one day whoever it is will get karma.

Just as we put the last bag down Lewis arrives and opens the coal bunker and we fill it for him. Lewis has a double hernia and is waiting to go into hospital in December and then he is going to recuperate at his ex-wife's house. That means lifting anything is out of bounds for the near future and so is getting stressed about someone trying to steal coal I tell him, but I don't think he hears me.

MAGGIE:

Next day we head back to Baltasound for milk. I asked yesterday for a sign in the clouds. I have had this a couple of times before where the clouds give me a picture sign and so has Pete. I was sitting at Haroldswick and right before my eyes a koala formed in the clouds. Interesting a koala. Does that mean we will go back to Australia?

On the way back, it was dark and Pete said he was bursting for a pee. We drove on a bit and he said he had to jump out. We were in the middle of nowhere on our way back to Hermaness and I stopped randomly. He stumbled on a keyring in the pitch dark and couldn't see what it was. I thought here he goes with one of his mojos-it will probably be something to do with an orca. No. It was a map of Australia keyring, in the middle of nowhere, in Unst. With that and seeing koala cloud earlier, I take it as a sign that we are going back to Australia one day. Yeee Haaa.

It's time to get a haircut. We had put an appointment on at the hairdresser. She is a Shetlander and I immediately like her. She tells us about her last excursion to Glasgow to see Kiefer Sutherland and his band play. She must really like him. It is such a big effort to go and see anything on the mainland. I think that is one of the things I would miss. The not being able to drive forty minutes to Glasgow to see a concert or theatre performance at short notice. That and seeing my family. She gives Pete the best haircut as most of the ends haven't recovered from the spider bite in New Zealand. My hair feels good too. I have had a lot of positive feedback on the colour of my natural grey hair. I let the natural colour grow in whilst touring Australia and New Zealand. We are going to an auction night tonight and we are not quite sure what to expect. I have been at a couple of upmarket charity ones in Ayr which I found quite boring.

PETE

We got up to another nice day. We've been lucky with the weather up here most of the time. We were in a shop and saw a poster which said All Welcome at Halloween Night—dook for apples plus Auction. Village Hall 7.30 pm.

We decide to go. When we reach the hall and go in, I think I recognise Mungo, Lewis' pal. He's busy talking. I see only one wean and she is on her way out. Sadly we've missed the dookin for apples. There's tables and tables of bric-a-brac. Everything is a pound or less according to a big notice. Great. My kind of bargain hunt. First thing I notice is Ravi Shankar CD for 50p. That'll do for me. Next I see a pyramid puzzle for 20p. As usual up here there are cakes in abundance.

After a while, chairs are put out in rows and a man gets up on the stage and shouts. GOOD EVENING EVERYBODY. Hope you are all seated it's time for the Auction tonight. First up are these, he holds up a box. Looking all around he says what do I hear for this lovely half dozen eggs. Do I hear two pounds. Three pounds. Come now they might be double yokes! Four pounds that's better. He lifts his hand up again and shouts at us all. LOOK THEY ARE EVEN IN A WEE BOX DO I HEAR FIVE POUNDS? Six pounds, seven pounds. They're big eggs. Eight pounds. Sold for eight pounds.

Right, next up we have another half dozen eggs. I'll start with two pounds... A cabbage went for eight pounds as well. A plastic carrier bag of potatoes went for six pounds (Mind you a bag of chips is about six pounds up here so I would say that is a bargain) A bag of kindling sticks went for five pounds. There were lots of odd things up for auction. They all started at two pounds, worked their way up then back down, then back up again. At one point, he pointed and asked a guy if he was scratching his nose or bidding. I've got an itch was the reply. All said and done I am really baffled why someone would pay eight pounds for a cabbage and yet not bother with a perfectly working mountain bike.

The last lot up was the best though. The auctioneer pointed to a table and says, "what's this?"

Oh it's a fish tank; it's got stuff in it so you won't need to buy any nick knacks for it and heck it's got water in it as well, so all you need is the fish...being a fishing island I wasn't surprised it didn't get picked up.

Actually I won a raffle ticket again and this time I won two dishcloths. I was just about to put them in the bin on the way out when Margaret spots me and says, "Hey we're keeping them. That'll do the house when we get one." A good night was had by all. We head off into the night hoping to see the Northern Lights.

MAGGIE

Next day we go out and about and end up at Northwick beach which has talcum rock used to make talcum powder. As per usual a couple of seals are watching us as we climb the rocks. They are so inquisitive.

Pete had been talking about his Granny McGillveray earlier. She had not spared the rod when he was a wee boy and had been sore on Pete and his two brothers. She was the Granny with the stick. He was talking about wanting to forgive her as she had a big effect on him.

We walked on up the rocks and came to a commemorative seat with a name plaque on it. Would you believe it the name McGillveray is on the plaque and there is a small rock sitting on the seat. On closer inspection, the rock has the name Pete on it and a poppy. The poppy is the symbol of peace and Armistice

Day is coming soon. He is always finding rocks but this kinda shook him to his roots.

PETE:

We decide to go to a small beach near the most northerly Church in Scotland. In fact, what that means is that it is the most northerly Church on the British isles and only a short boat ride from Norway. When we go in, we find out that it is twinned with the most southerly Church on the British isles at the tip of Lands' End and only a short boat ride away France. Seemingly the Parishioners visit each other regularly.

As we walk along the little beach way up here in the middle of nowhere I talked about my gran and grandpa. My grandpa was in the Navy and used to tell us stories about the Navy up here. My gran (nana) was called McGillveray and they didn't get on with other family members for reasons I am not sure of. I wonder if they are getting on now up above.

There is a bench sitting on top of the banking overlooking the sea and as we get closer there are two white doves holding olive branches and a red poppy engraved onto the cast iron bench. Sitting on the bench is a small white stone. When I look, I see it says Pete and there is also a little poppy painted on it as well. To top it all, attached to the back of the seat is a nameplate that says McGillveray. Well. I was shocked. Can you imagine going to sit down in the middle of nowhere and there is a remembrance seat and there in front of you is your own name with a poppy attached to it and the name of the grandparents you were just talking to your wife about? Hoping they are OK now AND there are two white doves holding olive branches. Fuck sake you just couldn't make this up. It is a strange and magical world we live in and we talk for ages about what's happened and the strange goings-on.

Ghosts and Laughing Gnomes

MAGGIE:

Next day we visit the Final Checkout cafe and have a soup and a baked potato. There is a gentleman at the next table. The gentleman lives on the Island of Fetlar. We tell him of our shock at the eating habits of otters. We tell him about the New Zealand possums. He tells us that once he was at the Ferry terminal on the island and a bull orca came racing after a seal and ended up on the shore just a few feet away from him. Now that is memorable! He says he has planted two hundred trees on Fetlar in the hope they will grow and plans to plant another two hundred more. These are the encounters I enjoy so much.

Next day it's a very, very wild. The winds are forty miles an hour plus so not wise to drive in it. Pete wants to go to the shop and decides to walk. It certainly will clear his head today. I stay and finish reading then do some writing. It takes Pete three hours to go and come back in the wild weather. Luckily for him there was no rain. We end up having to sleep in the wee howf as the van is rocking so much. The Icelandic wool blankets are actually one of the best things I have ever bought. We sleep sitting upright on the couch and I am surprised that I manage to sleep for eight hours with legs up on chairs, snug under the covers. It is cold outside but it's also a very, very wild night.

PETE:

Well today, according to the weather it's to be forty to fifty mph winds. It's single track roads so we won't be taking the van anywhere today because with the roof rack giving the van extra height we could be blown off a cliff and that's the last thing we need. Margaret decides she is going nowhere and thinks I am bonkers to go a walk. I think it's a great idea. It will blow the cobwebs away. I give Mags a kiss on the cheek and tell her I'll see her later and as I walk away I hear the words Mad…friggin mad…blowing in the wind.

I walk a good few miles. Sure enough it is windy, but it's not cold. Not long after I left I stood at a spot and to take in the view. There was nothing for miles. Then a man in a red jacket came round the corner and said hello as he walked past. I said hello back and walked on. I looked back to see if it was a ghost and thankfully saw him walking up the hill. I think to myself that he might be a twitcher, there's plenty of them about here, but he didn't have any binoculars that I could see.

As I walked on it might have been windy, but man it felt good. This place is so remote. Pure peace. As I walked further I saw the injured gannet we tried to help a couple of days ago. It wouldn't let us near it. This time, sadly, as I approached I could see it was dead. There are no predators in Shetland so I'm guessing it was exhausted or something like that. Life can be taken away in a second I think to myself, so once again I thank my lucky stars that Margaret and I can travel and are both healthy.

I reach the spot where the Viking Longhouse is and immediately think of the German girl who slept the night in it. She must have loved that. I also think of when we made our dinner in it and ate it round the old fire pit. I remember trying to carve a rune stone for Margaret. I remember smiling to myself sitting there with my beard and all feeling real Vikingy when a tourist walked in and looked real shocked. They must have thought we were some kind of re-enactment group because they turned round and walked right out as quick as they came in. As I walk on I pass an old quarry with a stone crusher machine sitting with the keys still in it. If I had to start it, I would probably wake up the full Island and the Wifey Ghost as well so I decide against it.

As I turn the corner, down the road is the shop called, no surprise, Final Checkout. I discover it is getting a lot windier and I am surprised to find out I have been walking for an hour and a half. The Final Checkout, a few houses that is Haroldswick, Lewis' little pad and the Muckle Flugga Lighthouse, then it's Norway.

In I go and I feel a bit pleased when the staff recognise me. I decide to buy a four pack of Guinness and put them in my rucksack, have a wee heat and then, well, head back. As I walk outside I find out that the winds have risen yet again and it's getting dark. I know if I stick to the single track road and don't go down any side roads I will be fine.

Someone offers me a lift as I am leaving the shop and I say no thanks. This is to show the hardy islanders up here that some of us lowlanders are hardy as well. They just look at me as if I'm not so much hardy as foolhardy to walk ANYWHERE in this.

I walk on and it's not long until everything starts to look different in the twilight. The old ruins look very ancient, all the rocks seem to suddenly have faces in the moonlight. Why didn't I accept the lift? This is spooky as I pass the Viking Boat and it looks stunning in the shadows. I think to myself that German hitch hiker girl is one tough cookie.

The wind is louder and the remoteness seems very remote. I keep thinking I'm seeing things and hearing things as it starts to turn pitch black except for the glow of the moonlight. I remind myself that there are no bears up here, when suddenly I feel my foot on something, Christ what is that? I look down and realise that it is Shetland Pony shit. I am so glad as it means I am on the right road. I have not wandered off it. I see a couple of figures coming towards me making a hell of a noise with their footsteps on the road. I hope I am not going to get jumped by Zombies.

As they get closer I realise it's the wee ponies with their shaggy hairdos and there's more of them coming now. All I can hope is that they are as friendly in the night time as they are in the day time. I nervously put my hand out to stroke the first one to arrive and say in a soft voice, "Hello wee chappie," it's all I can think of. As they all come up to me they all get stroked and called wee chappie. They walk down the road with me and I feel very safe. However, it's not long until they disappear like elephants in the jungle. I see a light in the distance and I know it's the remote farm and I've not far to go.

My eyes are getting accustomed to the light. I can see things a bit clearer now. However, as I turn the corner to walk up the hill, the moon disappears and it's back to the twilight zone. I can make out a sign and then I literally stop dead in my tracks and fk me, there in front of me in the twilight there are two fkn gnomes with beards, hats, laughing. For a moment, I stop breathing then I thankfully remember seeing them when we first arrived. Someone had the bright idea of cementing them to the side of the road and painting them. I am seeing them in the dark for the first time. I make a mental note to ask Margaret to go a walk tomorrow night, let her go first, not say anything. I see our van ahead and I'm glad. It's been an eventful walk. I can't stop thinking of those gnomes.

I am also beginning to think that it's so windy now that there might be a bit of truth in Lewis' tale about the wind being that fierce that it blew a guy's mouth open and he couldn't get it to shut till the wind calmed down.

MAGGIE:
Meant to go to Church today but our battery needed charged and we had to ask Lewis if he could jump start us. Too late for Church so not meant to be, but we go out and give Gabie a good run.

The wee Methodist Church here is the furthest north in Britain. We had a look inside the other day and it looks a quite Scandinavian design. I haven't been to a Church service since New Zealand. I haven't really had the notion as it all seems a bit dreich over in the UK and I really believe our Church is wherever we want it to be. I am a big fan of the band Faithless.

When we came back, Lewis had a Lyon coffee and caramel wafer waiting on us and I had a power shower. The coffee and the shower were both wonderful. We are heading back out to see the Fireworks display at the school. I hadn't put enough layers on and was glad to get back to the van. It only lasted ten minutes, but we saw a few folk we know.

Back at the van the winds are picking up again and the van is rocking so we decide to sleep in the howf again. After an hour or so, I heard a noise as if something is throwing rocks around or moving them. I then realise straight away it's rats. I recognise the noise as I have encountered them before when I lived at the estate in Scotland. I have an irrational fear of rats. I know they are not really wanting to harm me but they sound as if they are in the room. I shout FUCK OFF to let them know I am here. They stop for half an hour, then they are noisier than before. I feel the irrational fear creeping up my body to the back of my head. My subconscious. Maybe it's their association with the plague and disease. If they were Quokkas, the wee marsupials I saw in the Billy Connolly Australia book, when first spotted they were mistaken for rats. Maybe if it was them I would be smiling and wanting to go out and see them.

Rats are from the dark side. They make me face my darkest fears. I don't think people realise just how many rats there are in this country. They are so good at hiding, but they are literally everywhere. I also think that the litter louts in this country are playing to these creatures strengths as they are born survivors. Do they actually realise they are encouraging them by inadvertently feeding them? It might just come and bite then on the proverbial arse as the more of them there are the more chance of disease spreading.

We went back over to the van at 2 am to get peace from the rats. The storm had passed and the stars were out in force. It has turned out such a clear night. No sign of the Mirrie Dancers though. Had a great sleep in Gabie with our covers and when we got up it was a beautiful day with no wind whatsoever. There has been a first fall of snow. We went a walk to the beach as it was glorious AND we saw a rainbow.

PETE:
Today we were going to go to the Church then the shop, but not straight away because the battery went dead. The lights have been left on too long. I head up to see if Lewis the Lighthouse Keeper is in to get a jump start.

I walked up the hill, up the steps, chapped his door. There was no one in, but then he appeared out of his hut with his headlight on. He had been working on something. I explained what was wrong and he said not a problem. When he put his van next to ours, he brought out the longest jump lead I have ever seen. Lewis, you see also worked at the airport and helicopter pad. I think he must've got them there.

Lewis told us that when he was joining the Lighthouse one of the first things he had to do was when he reached the top of the steps he had to climb out onto the dome roof and check it at the very top. No safety net or anything. If you fell, you landed on the rocks. This, he said, was to see if you were OK with heights. Erm…no thanks!

Once the van was charged and we take it a spin, when we got back we got invited in for a coffee. The wonderful coffee bags. Just like tea bags, but bigger. That and a biscuit or two and Lewis was on a roll with another story. He showed us some of his photos. They are great: aurora borealis, birds, whales, scenery and also his family. We get the feeling that he is well liked. As we are leaving we get handed some giant coffee bags, smoked salmon, fresh salmon. On the way out, Margaret asks what's that on the wall? Sit back down and I'll show you says Lewis. My ex-wife thought it was a hat.

It does look like a hat, but on closer inspection it has the shape of a dome with a smaller round dome shape on top of it. It was thick. Thick as a car tyre. Hard and heavy. You could knock it as if knocking on a door and when you turned it over there was a strap made of some stuff across the opening along the back. Lewis clasping the strap at the back announces that it's made of rhinoceros skin and is tough as old boots and, according to what he has been told, it's at least three hundred years old. Lewis then announces it comes with these and he produces two sheaths the shape of quarter moons. He also brings out a sword which although not as curved has the same intricate design. Take the sword out he offers.

We have just watched The Last Samurai DVD and I just happen to be reading a book about a young Samurai so now I am getting the chance to take the sword out the way a Samurai would do it, although no one else in the room notices. They are made from silver and come from Oman. What they have rhinoceros in the Oman? Hell yeah says Lewis there would have been Rhinos in Scotland when the world was all joined together.

They are like something out of the dinosaur age and I wish they were still here. Imagine going a walk and seeing a Rhino. Now that's cool, but for now I will make do with Shetland Ponies and otters. Hmmmm I think to myself they do look Persian or Moorish and I remember seeing a film with Sydney Poitier in it and he had a curved sword… we have to go now so we bow and say cheerio to our host and thank him for his hospitality…the shop closes shortly and there is a firework display we want to see.

After getting some provisions, we drive to Unst school where the display is held and wait for everyone to arrive. Well there were a lot more children and parents but it's changed days, very sad I think. Health and Safety are taking the fun out of fun. There were no sparklers allowed, the fireworks were in a field we weren't allowed in and there was a Fire Engine. No fire, just rockets which took two firemen to launch and two firemen with hoses just in case.

It was over in ten minutes. The children though were having a ball. Every time there was a swoosh they would jump up, cheer and clap. Sweeties were being handed round. I was a bit disappointed. Probably, because we were up so far and because of Up Hella A, I expected something like a Brazilian Carnival or something. Better safe than sorry I suppose, at least the children seemed to enjoy it.

All That Glitters Is Not Gold

PETE:

We woke up and it's lovely morning so we decide to go a walk to the small beach again. It's a marvellous place and I am sure tourists would love it if they knew it was here. Whenever the sun shines on the rocks it glitters like gold or silver or both. You can't help but take a piece. There is proof that there are gold and silver here but not enough to cause a Gold Rush, so don't look out your pan yet. There's definitely not enough to start a mine. There is also specks of platinum been recorded, but I don't know what it looks like. I could have walked past it for all I know.

I put a piece of 'gold' in my pocket and walk along admiring the view. Seemingly the small rivers that flow from the top of the hill and weave their way down, leave small minute traces on the rocks and stones. The water is crystal clear and it sparkles all the time. River Gold I call it and I have quite a few pieces by now. As you all probably know it's pyrite or fool's gold, but I'm keeping it anyway and I will make the kids wide eyed thinking it's real.

Something happened that I need to ask Lewis when I see him again. Last night in the van a light kept sweeping past now and again across our window, then it would stop, then start again, then stop. I decide to go and investigate.

I walk along the small path in the dark and I hear voices where the lights are still coming on in the field above. I stand behind the wall and a big light sweeps across the field as if it is searching for something. I duck down behind the rocks and the wall to watch. I see a couple of men with torches and ropes. Now I am getting a bit scared. It certainly looks a bit unnerving and dodgy to me. There's another two men with ropes. All four men meet at the top of the hill and throw something heavy down to where the ropes are attached, then pull the rope tight. As we are as far north as you can get and very remote, what on earth are they doing? They weren't making a sound. I was afraid to cough in case they heard me. Maybe they were burying someone? That would account for the ropes getting tight as they lowered the body in the pitch black. Then two of the men knelt down. *That's it*, I thought, they are actually burying the body. After what felt like an age, they hauled the rope back up the hill. They got into the van and away they went.

Next morning I walk up to where they were to see if I could see anything in the daylight. Nothing!. Then I saw a piece of paper. I picked it up and noticed it was an Ordnance Survey Map of the exact area we are at. I could see no blood or anything else as I walked round. Sure is a mystery.

Then I remembered seeing a poster in the shop Rams £236. *Of course*, I thought, *how daft of me to think of a body*. They were nicking sheep. That would be what the ropes were for and that's why they were kneeling. They were tying their feet. That's what they were hauling up the ropes. This has got to be the ideal place for sheep rustling here in the far reaches where the next stop is Norway. I decide to ask Lewis if a lot of sheep go missing up here.

MAGGIE:
We got caught in a sleet shower and Lewis again invited us in for a coffee. Pete already had a feeling Lewis was going to invite us up. We all had a good chinwag. We have a lot in common and his stories are great. He gave us Amadeus and The Old Grey Whistle Test DVDs and we gave him the Shawshank Redemption and Doon the Watter. He let us hear some classical music. He likes nothing better than to put the lights out and listen to Mozart or an Opera. He was married to a Londoner, but when they separated they remained very good friends. When he gets his hernia operation, he will recuperate with his ex-wife. Lewis has very kindly invited us up for lunch before we head back to Lerwick and Bigton Fireworks Display.

We tell Lewis we thought folk were rustling sheep because Pete saw folk with ropes and he set our minds at rest. We both burst out laughing when we find out it was the local Coastguards doing weekly training at night time. He told us of the time a spaniel had to be rescued as it had fallen eighty feet down the cliff whilst chasing a rabbit. He said it didn't stop licking the coastguards face as it got rescued.

We head down to the campsite at Yell to do some laundry and clean our van. Then it's to the ferry around 4 pm. We call in at Brae for a fish supper and at last we got an excellent fish supper on Shetland. They have a great choice such as crab patties and all sorts of different fish.

We get to Bigton quite late and I see a huge fire blazing in the distance. I don't like being late and try to keep calm about missing the start because I have learned that getting wound up won't change anything. I manage to regain my composure quickly because I recognised the trigger. That is progress. No blame here mate.

Turns out all the locals are just arriving as we are. There is a rope barrier for health and safety reasons, so we can't feel the heat. It was a great fireworks display and there is hot chocolate for everyone.

Such great community spirit. It's a lovely clear night but we need to be wrapped up. Couldn't believe my ears when I guy called out for his son Titus! That's a lot to live up to! Once everyone was away to the hall for burgers we got next to the fire. It was a beauty with purple, blue, green, orange and yellow flames. We so love a bonfire. We stand for an hour and then head up and have Sausermaet and onion rolls. We ended up staying to watch The Taste of Shetland cookery competition Muckle Bites for the adults and Peerie Bites for the under 18s.

We got to taste some seaweed fed lamb and it was delicious. We were pleased that we met our friend from Maybole on the way out. We head to the fire and it was still blazing with lots of embers. After a wee, while we got into the front of Gabie and watched from there. We were hopeful that we might see the Mirrie Dancers as the young guys clearing up said that they had often seen them here. There is a bit of luck involved I suppose.

PETE:
Well, that's Saturday here and off we went to the bonfire. I am disappointed that they have a big rope cordoning off the crowd from the bonfire. At least, the

bairns have sparklers. It's so far from the fire I'm frozen. The guy lighting the sparklers is holding a Bunsen burner with big fat gloves. The two guys that let off the fireworks are dressed up to the hilt. They even have firemen helmets on with their thick, padded jackets and are a field away. The fire was huge with blue, green and yellow flames lighting up the surrounding beach and rocks. The big moon helped to add to the atmosphere, but it didn't take away the fact that we were that far away from the fire there was no heat. What a waste of a belter of a fire.

We managed to grab a cup of hot chocolate each which tasted great and was warm! We were told that the Hall was open and everyone was welcome to come for burgers and there is a bar. As soon as Margaret heard there was food I knew we were going.

We waited till everyone had left and then went over to the fire and got a heat. We stayed till the flames went down and then, as we always do with fires big or small, we looked at the embers for faces, maps, animals etc. Great fun and it doesn't cost a penny.

Time to head to the pub and food. When we got there, we realised there was a cookery competition. One for the young ones (peeries)and one for the adults (muckles).

Not being ignorant I supported the cause by buying a pint and Margaret supported the cause by buying some burgers. Results came through the winning Peerie was a fish dish and the winning Muckle was a lamb dish (remember the seaweed lamb) made with a local gin sauce and a cheesecake with a bramble and gin sauce. I made sure we got to taste that and it was absolutely, deliciously, scrumptious.

Margaret thought she recognised someone and sure enough it was the winner's sister and she recognised us and asked about the whales and showed us footage. We also met our friend who lived in Ayrshire. A good night was had by all and we went back to the dying embers to look again for anything that sticks out. We end up camping just along from it as that is the beauty of camping.

MAGGIE:

Next morning it is beautiful weather and we go for a walk on St Ninians Beach one of our favourite places on Shetland. Then it's off to Lerwick.

We are back at the library. I have been reading Michelle Obama's autobiography. She is a remarkable lady. So inspiring and she believes in education the way that I do. I feel that Scotland, as a nation, has lost its appreciation of education lifting you up. We used to lead the way. We believed in education for all long before most of the world. Now that is true socialism. Look at all the things we have invented.

I am also reading Eat Pray Love by Elizabeth Gilbert and also her book Big Magic on creativity, both of which I am enjoying immensely. She writes about a poet and for some reason it reminds me of the name of the soldier who had a big effect on Pete in Little Akaloa New Zealand and who died on 6th May 1943. I look him up and find out he died of his wounds in Tunisia where he is buried in a war graveyard. I hadn't realised it was 11th November, until the site I was on offered to lay a poppy for him. I laid a virtual poppy for him and prayed for peace on earth.

PETE:

We wake up the next day and the fire is still smouldering. We stand a while feeling the heat and looking over the water. It's like a mirror, so we decide to head to the St Ninians Beach again. It's beautiful so we decide to stay and go a walk. I took a photo and when I looked at it my shadow was huge and looks like a giant hare. Love it. Meanwhile, anything to do with birds and Margaret is right in there and sure enough she is away at the far end of the walkway checking out a bird she recognises.

We move on to Lerwick as it's late and can't resist going to the Broch. After all, it is 11th of the 11th Armistice Day tomorrow. I decide to take with me the stone I found. I keep finding them even in other countries. They've usually got a message on them and then you pass them on somewhere else. I place mine in the Broch for someone else to find. For the record, it says Hopeman Rocks.

As usual the Broch zaps my phone. I tell you there is some energy in there. Anyway, it's time to head back up to Unst, so we go to Yell first and camp at the site where the roof is a boat. We do nothing all day except clean the van, do some washing and look for the Mirrie Dancers…still no orcas.

It seems no time we have to collect our stuff, say cheerio to our wee writing place and our favourite Lighthouse keeper and friend Lewis.

Afterwards we call in to visit Muness Castle. I go in and wander about, but Margaret says she is not going in as it belonged to the Stewart brothers who were a real nasty lot. They killed and maimed a lot of people to get their castle so no way is she even going near it she says. I try telling her it was hundreds of years ago, but she says bad vibes last longer and she is not going in. Fair enough. I like old castles and as long as there are no Ringwraiths hanging around I'm OK with that.

After the castle and since it is not windy, we decide to camp at the Youth Hostel one last time. When we wake up, it's another nice day.

As we drive off and reverse a back wheel ends up in a ditch. Luckily it has landed on a traffic cone in the ditch so it is not too bad, but we are going to need help to get out. We remember that one of the friendly women from the Makkin Belts night stays nearby and she helps us by phoning someone who very kindly pulls us out with his 4 × 4 and a rope which he tells us to keep. That's a favour we owe you. Many thanks.

With that we head to Burravoe where we give everything a good clean and airing and have an early night to try and see the Mirrie Dancers.

MAGGIE

It's time to say cheerio and thank you to Lewis. He was been so kind to us and also has kept us entertained with his stories. We are both sure we will meet him again sometime. One day in the future I would love to book a holiday nearby. Before we left, he showed us a photo of Tonga stack, a piece of flint in the shape of a seal on one side and polar bear on the other and a lovely piece of driftwood. He is our kind of guy.

We called in to see Muness Castle, but when I found out it was the Stewart's castle I wanted nothing to do with it. We had asked about my maiden name Dalziel connection to Yell. There are quite a few Dalziels in Shetland. Yikes I found out there was nothing until the 16th century so that rules out Vikings. There are Lanarkshire connections to the name but I thought it went further back up here. There was Robert Stewart who became 1st Earl of Orkney and Lord of Zetland (Shetland) who was a bastard son of King James V. He had a dark, tyrannical heart and ruled Orkney and Shetland with a rod of iron causing much suffering. His son Patrick followed suit (he was executed for treason in 1615) Could it be that some of the Lanarkshire Dalziels came up to Shetland in the 16th century with the dastardly Stewarts because Stewart was a half-brother to Mary Queen of Scots and the Dalziels had connections with Mary Queen of Scots? I hope that this is wrong. Mmmm needs further investigation. When I think of it, the Vikings were always on the rampage and probably slaughtered a lot of the

Picts so their history is a bit gruesome too. We could get our DNA tested methinks.

PETE:

No luck with the Mirrie Dancers last night although it was a clear and frosty night. The campsite is great here and it's all run on donations and by volunteers. Although it is cold outside, we are as warm as toast with our Icelandic wool blankets and lined curtains. If you have the right gear and are well prepared, camping in winter can be cool.

The only other person staying here knocked on our door last night and asked if we wanted a hot toddy as it was so cold. The offer of a hot toddy is not to be sneezed at so we gratefully accepted. He likes his rum and coffee to make his Shetland Toddy. They are great. He gave us a very good measure. Then as he was showing us his two Labradors' and their paws, which are webbed, that's why they are good swimmers, we got another good measure. By this time, the two large toddies were taking the shivers out of me timbers.

He was telling us that the woman he was working for had very large windows to see the Northern Lights. None of could work out why she didn't just go outside to see them and save a fortune. Ah well I suppose that's what rum does to you...he told us Shetland says Dat for that so with dat we said thanks and moved on the next day to get the Ferry back to Lerwick.

MAGGIE

We stop at Yell for at least one night before heading to the mainland. We got talking to a guy who has been staying here long term. He is a stonemason and he is here for as long as the job lasts. He has two lovely black Labradors with him in his caravan. He invites us in for what he calls a Rum Toddy. He introduces himself and I love his accent.

Do you want coffee in your rum toddy? no thanks-sugar? no thanks—milk? no thanks. He then proceeds to hand it over complete with rum, coffee, sugar and milk. Much to my surprise it is delicious. Rum is not usually a drink I would chose but he has converted me. He says it is a local favourite. Cheers mate. Pete has lots to talk about as he used to work for a stonemason. One thing I remember is him saying he had a mouse in his van and he had set a mousetrap which caught it last night.

Next day we stay quite late and then decide to go to the mainland as we are having no luck with the Northern Lights, but the weather has been fabulous for the time of year. No wind. Frosty. Healthy.

A guy we got talking to told us he had seen some orcas close up in Yell sound. A wee lassie on the same trip was screaming with delight. These are things that we all remember at the end of the day.

Once back on the mainland we have a phone call to make as we get a better signal. We said we are in Brae and she said that is perfect as she is coming up to Hillswick tomorrow to see a photograph exhibition. Perfect. We are going to meet her after 4.30 tomorrow.

It's another beautiful day in Brae and heading to Hillswick later. We end up falling in love with the place. We go to the Shetland Photo Exhibition and I think of Lewis' photos. It is great seeing the photos conjuring up a different time and era. We enjoy it because we can vividly remember these times and now they are history.

We are waiting to meet Christina who we found out recently actually owns the wee storehouse as it did not belong to the RNLI as we had thought. We are meeting her to make a donation to the RNLI as a thanks for the use of the wee writing howf. Looking forward to meeting this woman as she sounds lovely.

Once we say our goodbyes to Christina we buy a couple of White Wife Ales which I think tastes like it was made a couple of hundred years ago. They guy has seemingly gone out of business. They are definitely an acquired taste. Speaking of white wives. We also find out that there is also another famous White Wife and not just the ghost. She is at Otterswick and is a reconstructed figurehead from a German Sail Training Boat which in 1924 was on the way to Chile. The captain of the ship made an error, the ship crashed and sank here, on the coast of Yell. The ship was called Bohus and the figurehead looks out to sea in hope and as a memorial to the four men who died and are buried at Mid Yell Cemetery.

We manage to have our breakfast outside next to the sea as it is a lovely morning. We get to speak to the guy who runs the Wildlife Sanctuary. There is only one seal pup at the moment and he is called Xander. We find out that came up voluntarily on the beach and was rescued by a local lady who was out walking. He was very thin and would have died if the sanctuary were not there.

We got to meet the lady who rescued him as she walked up the beach and we got talking. They also have a couple of otters, but we are not allowed to see them as they are being prepared for the wild and we would leave a human imprint. They need to think wild. He said a big change is gonna come with regard to Planet Earth and climate change because it has to. I said to him about me being in Friends of Earth in the 1980s and although being made a fool of then, it is all coming to pass. I too believe a change is gonna come. Don't know in what form though as Mother Nature is the boss and always finds a way to find balance.

Last night we headed down to the St Magnus Hotel in Hillswick. We have had no TV for the past year and we happened to hear that there is a programme on BBC by David Attenborough about Australia. We hoped that we might see it. At first, the signal was poor, but we kept the faith and sure enough when we flicked through all the channels we got BBC London which was a great signal. The kind lady behind the bar put subtitles on for us and the guy put the sound on as well. So kind of them. I had a coffee with a Baileys in and Pete had a pint of Lerwick IPA then a Guinness each. We enjoyed the programme so much. It is such a pleasure to hear David Attenborough's soothing voice and it was so exciting to see loads of the Australian wildlife that we had actually seen in person. Especially our good friends, the fruit bats.

PETE:

As we were driving along from the ferry to Lerwick the scenery is stunning. We decide to camp at Brae for the night where you get the best fish supper in the world apart from a Barramundi in Australia.

When we wake up, it's another sunny day and we decide to head to Hillswick. It's like a scene from Tarka the Otter. A lovely wee place with a shingle beach and Hotel up on the hill. We are heading up to the Hotel as David Attenborough is on tonight and it's about Australia. The owner allows us to put the subtitles on and watch it in the lounge. It's a real treat and we thoroughly enjoy seeing Australia again.

Next day we are going to visit the Wildlife Rescue Centre. We get to see a baby seal pup rescued a few weeks ago and hopefully going back into the wild. The garden has a ring of engraved standing stones with a Cairn of Hope. We walk along the beach in the hope of finding a suitable stone to add to it. I also put a Shetland stone that I found as it's time to pass it on. It's all good fun.

I found a piece of flint on the beach. I know it is flint because Lewis showed me what it looks like. His is a belter. It looks like a seal on one side and when you turn it around it looks like a polar bear. Mine looks like a pebble. I was hoping to find a piece of driftwood. Some of them have great shapes. Lewis has one that looks like a horse's head and he's going to wax it.

We also go into a studio showing an Exhibition of photographs from the 1970s. Love it. A strange thing happened in the chip shop. There was a guy standing waiting and started whistling Kate Bush Wuthering Heights. You know the one. I tried it and I can't do it. Try it and I bet you find it's hard to do. The other thing that caught my attention was a guy had a skateboard wrapped up in brown paper with postage stamps on it and the wheels sticking out. I could just picture the recipient receiving it and saying I wonder what this is?

MAGGIE:

We head back down the road to Lerwick before it gets dark and stop in at Brae for another chippy and to see if they had any crab cakes but they were sold out again. We see they have Ling on the menu and the girl tries to describe it and says it is kinda 'meaty'. I see what she means. It is a fish caught up here for centuries and although I prefer haddock, I am glad I tried it.

Pete decided to go to the Library, but since it is such a lovely day I chose to give the Broch a visit. I decided to walk round the Broch three times and then go inside just to enjoy the peace. After about ten minutes, a guy around ten years older than myself comes in and introduces himself as Henry. He asks me what I like about the Broch as he is local and comes here often. I try to explain that I feel close to my ancestors here and it has a great energy.

It turns out he has travelled the world. As he never married and never had children he feels that this allowed him to go anywhere he wished. He has no regrets he says. I say no wonder as he seems to have been in every corner of the globe. He has been to Israel twenty three times. He has walked the road St Paul took in Turkey and it meant a lot since he has a strong faith. He has been all over America and Canada, cruised the Greek Islands and been to Australia and New Zealand. He said he loved the penguins in Australia. I said I didn't realise Australia had penguins because it's so hot. He was a lovely gentleman who seemed very happy with his lot. I told him about Pete and that we are hoping to see the Northern Lights. He said you need a bit of luck. As he left he said give my regards to your husband because you are so lucky to be able to travel with someone who shares the same interests as you. I agree.

I met up with Pete at the library and we are thinking of going to an author talk there, but then Pete remembers that the Mareel has a film on the cinema called Wilderfest and something tells me to listen to his intuition. Pete is very in tune and I trust his intuition.

PETE:

In Lerwick, I decide to go back to the Library. I found out that Treasure Island is supposed to be based on the Isle of Unst and decided I want to read it again. I haven't read it in years so it will be fun to picture it, this time round, on Unst. I also got Kidnapped as if you remember it is all about pirates and caves as well. Putting me in the notion for a oho ho and a bottle of rum.

When in the Library, a young woman who works there came up and said. You were on the big screen yesterday. She went on to explain that last month after I agreed to get filmed they liked what I said and showed it to the Big Wig Council folk and it got five out of six votes. This helped to secure funding and keep libraries open. I am well chuffed.

As a treat we go to the Mareel to see a wildlife film I remember seeing advertised on a poster. It's actually lots of small films taken all over the world with two guys coming on stage in between each film to talk to the audience. After seeing it, I say to myself. For God's Sake, people wake up and stop destroying our home. We are in all it together.

The Mareel has a bar/cafe area so I get a pint and Margaret opts for a Moroccan mint tea which I can smell a mile away and reminds me of my trip to Marrakesh. We remember we have a DVD borrowed from the library. It's called Tolkien. I feel like I have had a lot of screens today and so with that it's good night... PS still no orcas.

MAGGIE:

We arrive in time to find out it is in the Auditorium and it is a kind of educational, awareness raising, charity event. There are lots of short films from all over the world. Coincidently, the first film is about penguins in South Africa and I immediately think of Henry talking about the penguins in Australia.

The event was very interesting, but also very sad and moved me to tears. There were a few short films such as whales and how humans are making the ocean too noisy and the whales are struggling with their sonar. Also about curlews disappearing in most of the mainland. I miss the Peesies (lapwings) in Ayrshire. There were thousands when I was young and now there is only a handful if you are lucky. It's the old story of you don't know how much you miss something until it is gone. The whaups (curlews) are going the same way and my brother is involved in a conservation project very similar to the one I watched here. In fact, he won an award while we were up here and he had to go to a Ceremony and it got presented to him by Kate Humble.

PETE:

We are in a shop and Margaret remembers she needs a first class stamp.

The guy says sure, do you know how much they are?

Not a scoob says Margaret.

I'll look it up it says the guy. He gets his phone out, laughs and says this will probably be sixty degrees north prices.

No signal nothing is happening. Just give me 50p.

Done says Margaret and hands it over swapping it for the stamp quicker than a bat out of hell. She likes a bargain does our Margaret.

MAGGIE:

Next day Pete and me had a heart to heart about the power of the mind and hypnosis and abuse of that power. Subliminal suggestion and brain washing. I saw my ex-husband hypnotised at a show in 1981 and he did things that he wouldn't have dreamed of doing as he was quite shy. It was very entertaining and it opened my eyes to the power of suggestion. A change has got to come.

Here I am sitting in Shetland library with my diary. Pete's wee video got shown to the bigwigs and Shetland Library got a 5/6. Well it gets a 10/10 from me.

We heard that the Mirrie Dancers were seen on Unst and that the whales were seen at Burravoe the two places we had just been and just left. Ah well it looks like we will need to come back up here one day as we have booked the ferry to sail back to Aberdeen and home on 8th December.

Feel I am ready to go home to see our families and to our new wee flat.

PETE:

Well! Let me tell you this. If any of you are thinking of doing what we have done in a campervan, then be warned. If you leave your doors open, then be afraid, be very afraid. We got a mouse last night or the night before, who knows when? We heard it at 2 o'clock this morning. It was too dark to do anything so I went to sleep. However, Margaret hasn't slept a wink so we drove to a campsite which has facilities. The van is empty and I mean empty. There is only a single mouse trap with a piece of cheese on it. The doors are all locked and since we are the only campers we are sleeping on the kitchen floor tonight. I am hoping that little Mickey is away tomorrow. Spiritual or otherwise. In the meantime, we learn from a local that there is a huge seal that comes up every now and then onto the plankway to the boats. Its name is Hughie and he doesn't move till he is fed.

We wake up on the floor in our makeshift bed and we've both slept very well. We have our coffee and breakfast and go out to check and see if little Mickey has had the sense to bugger off. I peek slowly through the curtain and there's little Mickey passed on to the great mouse hole in the sky. I tell Margaret that the van is now ready to put everything back in, but no, Margaret decides that there might be two mice, so we're on the kitchen floor tonight again.

MAGGIE:

We had to buy a mousetrap today as we keep hearing a scraping sound at night which is a bit creepy. The mind starts playing tricks and images of rats come into my mind.

We have decided to drive to the campsite at Cunningsburgh. We had to take everything out of the van and I mean everything. My God I realise that we have far too much stuff with us as we haven't used a lot of it. We had to set a trap as we have no other option. We don't enjoy doing it and we sleep on the floor of the kitchen in the campsite which we have to ourselves because of the time of year. I am relieved to see a mouse caught in the trap. I did shed a wee tear for it. I am not scared of mice, but they can do a lot of damage. Plus it's rummaging was keeping me awake. I have a feeling that there may be another mouse as the scraping was coming at two different areas at the same time, so we decide to stay for another three nights and keep the trap set in the van.

The campsite is ideal. We can sit and write all day as we have the place to ourselves. We go a walk, since it is a lovely calm day and good forecast. The shop is about a mile and a half away and so we walk there and have a lovely plate of soup.

Next day the weather forecast is correct and it has been wild so we decided to write all day. We got talking to a maintenance man about the mouse and he mentioned rats. I could see him shiver as he had worked on merchant cargo ships from South America and saw lots of them. Give me the shivers too. I can cope with mice, snakes and spiders but rats eeek!

PETE:

This place is called Aithsvoe Marina and I highly recommend it. It's a lovely spot and it's spacious with big windows looking out at the little boats bob, bob, bobbing in the harbour. Due to the fact it is November coming into December we get the place to ourselves and are able to write, take shelter, spot birds, otters and maybe even orcas.

We've been told that gales and rain forecast for tomorrow so we might not be going anywhere. I take a walk round the little inlet or voe and look at the names of the boats. I like to do that when I can. It's great fun as you come across some corkers. I've seen a Bilbo Baggins, a Sergeant Pepper, Aurora Borealis and in Australia a Brown Butt. Here though it is a little more homely with names like June Rose, Magdalene, Coastguard, Peggy Lee and there's also a Margaret.

I got talking to a guy who says there is a good chance of seeing the Mirrie Dancers tonight. He also told me he has seen them swirling around in a circle and that's why they are called dancers. He also points to a sign in a field which says Beware of the Ram and he says we're not kidding!

MAGGIE

Winter is setting in. The sun rises at 8.40 am and sets at 3.05 pm. The weather has changed. We are so glad to be in the campsite as the winds are wild. The locals are lovely and very welcoming and wish us luck with our writing. Pete found another one of the rocks and it said: If you are not smiling you are doing it wrong. I had literally just been writing about the two big Māori guys from Hokitika and their big smiles and big hearted generosity.

It was sixty mile an hour gales last night. Very stormy. I saw flashing lights and thought it was lightning at first then realised it was shooting stars lighting up the whole sky. It was like being in an observatory. I remembered the guy said yesterday if the sky is clear you can see the Northern lights even if it is blowing a gale, so here's hoping.

We were talking about the Shetland names up here. There is a definite Scandinavian flavour. There is Magnus, Malcolm, Helga and Lottie and we heard a dad call out to his wee boy Ragnar. That really pleased Pete as he loves Ragnar of the Hairybreeks.

I believe in Karma. Here is my theory about Britain. We went off and conquered lots of countries. We said to the indigenous peoples this is now ours, we know best and you lot better adopt our ways. Now we have immigrants coming in and we don't like it as they are imposing their ways on us. We are all one. We are not separate. The young people know this deep down. They have to go out and vote or the old farts will win.

From Air Balloons to Cars

PETE:

Good Morning Shetland. It is Tuesday 26th November 2019. We have both slept well, so well in fact, it is 11 am. No hurry as we are camping. You should try it sometime. I know it is not everybody's cup of tea and we got told we were daft to come up here in the winter months, but we love it. There are loads of places to visit, the folk are great and tonight we are going to the Pictures again. This time to see Aeronauts. It's seemingly based on a true story about way up past the clouds in a hot air balloon charting the weather conditions.

The Picture House (Mareel) is also a cafe/bar and that's where we end up before going to see Aeronauts. I think we have picked a chess night as there are loads of people sitting at tables playing chess. Some as young as eleven or twelve and others sixty or over. They've got stop clocks on the tables. I can play chess, but not good enough to put a clock on the table. It's time for the film and afterwards we'll see if we're in the van in Lerwick or Back at the Aithsvoe campsite.

MAGGIE:

At the book shop in Lerwick, I recognised the voice of a woman who was working there. It's the lady from Hillswick who found the baby seal. When I asked how it was, she tells me sadly it died. It wouldn't eat properly or go into the water pool at the Sanctuary. She said that friends were coming over from America to carry out an autopsy to find out what is going on. I was shocked to be told that Orca Whales in some seas are unable to have babies because of the levels of mercury in the water. I am happy to hear that the Shetland orcas are breeding well. If folk are coming all the way from America to do an autopsy on the baby seal, then alarm bells must be ringing.

We are going to the cinema tonight to see a film called The Aeronauts. I loved the Mareel Picture House. It's really an arts centre and the arts are really thriving up here. The Shetlanders wisely invest in their traditional music and culture and it really shows. We are making the most of the rest of our time here by going to see a few films and also a concert Carols by Candlelight, just before we leave.

Hughie and the Lights

PETE:
We decided to go back to the campsite for the sake of the weather and not having to drive in the morning. When we get up, it is a nice day. In fact, looking out the window I decide to make the most of it and go for a walk. I go to the bathroom, come back to the kitchen and it is raining. Where did that come from because a minute ago it was a great day. It's like that up here, it can become dark as quick as the flick of a light switch. I don't mean dusk I mean dark. Mind you dusk is at three in the afternoon at this time of year.

I decide to put my Wellies on and see if I can spot the giant seal I heard about. I get talking to a guy who tells me the seal's name is Hughie and he is huge so if you say hi to Hughie, a Huge hi becomes Hughie. Get it? Anagram. The guy also says it has to clear up tonight and we might see the Northern Lights if we look north. I say cheerio to the guy and go and look for Hi Huge. I also get a flask ready for later on as we try to see the Lights. I am excited and looking forward to seeing both.

From here, I can see Margaret sitting writing in the kitchen area through the glass window. I can't wait to tell her there is a chance of seeing the lights if we find a nice dark spot. According to a real life Aurora Hunter webpage he says the sun gave off a 4×6 cosmic shower which he saw through his filtered so and so telescope and therefore there is a good chance of seeing the lights tonight. We get into the van, flasks ready and also a blanket. We are prepared for a wait.

At first, it is starry, then we see a glow, but realise it is the town of Lerwick in the distance. We decide to look for another place to park and wait. We notice that it is getting cloudy again and we have ended up on a single track road. We hate reversing the van, it is much more difficult than a car. Visibility is difficult so we decide to reverse while there is a little bit of light. I go out to be stopper (I first heard this term in Australia and New Zealand). It is definitely very wise to have someone out when reversing the van to give a warning and slap the van when too close to something, No sooner am I out the van than it starts to snow. Margaret can't see a thing. I remind her it's one thump for stop and two to gently move. Margaret finally manages to turn the van, after a thirty two point turn and we head back to the warmth of our new found pad.

MAGGIE:

Time is running out to see the orcas and the Mirrie Dancers, but it's a good excuse to come back up here. I also want to see Up Helly A sometime. I am delighted to have seen otters, in the wild, up close and personal on more than one occasion. That is a dream come true and so many species of birds. It is twitchers paradise.

We are going to see Frozen 2 later. Disney is so uplifting and there is a lot of wisdom in the stories. Usually great songs too. Really looking forward to it. I absolutely love the first Frozen and even have the soundtrack to it. I was not disappointed. Disney does it's magic once again.

PETE:

Yesterday was a reminder to always be careful driving. We were on a single track road and as we turned a corner, we met a huge tractor nearly head on. It had two big forks jutting out the front of it. We were lucky in that the tractor was the one able to manoeuvre into the ditch type track on his side. Our mirrors got hit and they did their thing of flipping to the side. I suppose we were lucky to come out of it with just a slight graze and bump but paint still intact going by the size of the tractor. We both agree that he could have reversed back a bit. However, we both learned a lesson that there could be anything round a corner, even in the middle of nowhere.

Last night we went to the pictures again. They know our faces by now. We went to Frozen 2. Margaret was going so I could stay here if I wanted. Better Frozen 2 with Margaret, than frozen here myself. Let me tell you. I enjoyed it. We then had a chippy then drove back to our wee haven.

We had a bit of a sleepless night because of the high winds which were sixty miles an hour again. Tonight and the next three nights it has to be the same. We have moved back into the actual building instead of sleeping in the van. There will be no Northern Lights tonight I think and we have only five more nights left in Shetland. I saw another seal, but not Hughie. Before I go, there are three things I want to see: Hughie; Northern Lights; Orcas.

MAGGIE:

We went to see the film Last Christmas and we both really, really enjoyed it. The Leading actress was fabulous. She lit up the screen. It was great to see a film which deals with spirituality and energy, love and kindness.

I got my DNA result through but Pete's still to come. The result jumped up at me as I wanted to wait till Pete could see his too. I found out I am fifty per cent north west European, twenty seven per cent Scottish/Irish/Welsh and twenty two per cent English. Can't wait to see Pete's result.

It looks like the theory of being Scandinavian is out the window. I suppose north west Europe includes Denmark. I think it is because I feel an affinity to the way the Scandinavians live their lives now and their true socialism nowadays more than the pillaging of the bygone days. I know Scotland could have thrived and been governed successfully like Norway, Denmark, Sweden if we had just the right people in charge back in the day of the first Referendum in the 1970s.

Maybe it's being in the Broch so much, but I feel a real closeness to the Pictish people who lived here all those centuries ago. They seem to have been more peace loving and more advanced than first thought. In fact, I like to think that they were quite like Hobbits, small in stature and big in heart. The Shetland ponies and small Sheltie collies are the same.

Hughie – Orcas – Mirrie Dancers

PETE:

It's hard to get a signal up here and it's all run by volunteers and will cost around £15,000 for a line and a mast. I do hope they manage to raise the money somehow.

Last night we listened to the sixty mph winds in the comfort of our writing pad. It was noisy, blustery, heavy rain and winds but we were very cosy and comfortable and it was quite awesome to see the rain falling in the shadow of the lights. I even saw someone with oilskins working on a boat. It can be a hard life being on a boat.

When we wake up, it is a peaceful, calm morning and as the oilskin guy walked past me, I asked if everything was OK. Oh yes, I was just checking out the new anchors as we got them for all the boats. He said he was a volunteer. A couple drove up on a van last night, stayed in the carpark, only came in to use the showers. They complained about the prices. Well, I said, we think it's great and we don't mind putting the money in the envelope as you go out the door. It will help them get their phone mast…

We also went to see *Last Christmas*, a George Michael-themed Romcom. I was dreading it at first, but I am glad to report that I thoroughly enjoyed it.

As we were getting ready to go, Margaret came running in to say she can see something making big waves and ripples in the water. Could it be an Orca at last? Yeehaa there, gliding under the water was Big Hughie. He wasn't fully out of the water, but we could see just how big he was. We were so glad we got to leave on a high or should I say a Huge Hi…

MAGGIE:

Been writing all day and it's a beautiful calm day. No wind whatsoever.

Going to *Carols by Candlelight* tonight. We have our last supper. It's fish with red cabbage and roast potatoes, all very tasty.

What a great night to finish here. It's a full house and a great atmosphere. There are banjos mandolins, guitars and some bluegrass and it was all heartfelt. My favourite carol on the night? *When a Child Is Born*.

PETE

We finally managed to leave Aithsvoe and head to Lerwick for the twelve hour ferry journey across the North Sea. There is one last thing to see before we go. We're off to see Carols by Candlelight by the Ness Brothers.

Our last night saw us watching the Ness Brothers Christmas Carols Bluegrass style. It was Fab. Banjos, fiddles, violins, cello, piano, mandolin and of course, slide guitars. All that and mulled wine which felt a fitting end to our adventure. It made me feel like a real wee pict, I know they didn't have slide guitars back in the day but hey!

Aaaaand, guess what…

I got my DNA results just before the concert sitting in the Mareel bar…and I'm pleased to report that I'm 72% Celtic. In other words, I'm 72% Pictish. I knew it… I mean, let's face it, I'm small, hairy (ginger in my younger days, I may add), beardy and nice-natured. A bit like a Hobbit actually, so I'm well chuffed it also means I'm madefaegirders. GRRRRRR.

I'm only 20% English, well, I can live with that, unlike Margaret who's still reeling over the news she's a massive 50% north west European and only 27% Celtic, 22% English and here's her hoping she was a Viking. WELL! I did warn her what that bus driver told me about researching your family tree—be afraid, be very afraid...which reminds me: I'm 3% Finnish which makes me more Viking than Margaret.

Mind you, I'd much rather be Celtic/Pictish than Viking after reading up on it all. To top it off, I'm also 5% Italian...hmmmm. God knows where that came from. Margaret reckons it's from one of the Roman soldiers from the lost Legion.

It sure is a strange world and up here in the Shetlands for some reason, I feel right at home. Probably because I'm a Pict. I think I told you before that way back in the day, there used to be bears up here. They lived alongside Picts as friends till one day, a rogue bear actually ate a Pict. When the tribe realised what happened, they cut cords with the bears and that day became known as the day the teddy bears had their nicked Pict.

MAGGIE:

The trip from one end of the world to the other—there and back again—really has been an amazing adventure and we lived to tell the tale. We have lots of great photos and memories to look back on. This adventure allowed us to get up close and personal with some incredible wildlife, see some spectacular landscapes and encounter some wonderful people, with lots of synchronicity along the way. What I didn't realise, when we set off, was that the adventure would also allow us to get to know ourselves at a such a deep level. I reckon this is the greatest gift we can all give to ourselves.

Travelling really does open your mind. Living in a campervan showed us that we all need very little to live happily and in harmony. It also helps to appreciate the simple things in life. We felt lucky to be alive which in turn seemed to bring us luck. The joy of having nature surrounding us really was a tonic. We had no TV, no internet, no newspapers and a lot of the time no music, which we found out allowed us to connect to something much bigger and wiser. The peace and the quiet allowed me to get to know myself really well and it also allowed my commitment to spiritual growth to take place. To be able to keep calm and know that everything is taken care of is truly to have faith. My faith is growing stronger every day. I have come a long way in more ways than one.

PETE:

The Shetland Islands remind me of a bygone era of the Scotland I knew when I was growing up in the 1960s and 1970s. You could leave your jacket sitting anywhere. If you dropped your wallet, you could retrace your steps and probably find it. People were friendly and helpful and took responsibility for their actions. Nowadays, if you leave your jacket, consider it lost or in cash converters. If you drop your wallet, you find it empty and over a dyke.

Here in Shetland, the people are very friendly and go out of their way to help you. You could go and leave a pile of coins sitting and come back the following week and collect them untouched. The folk here all have got a story to tell you: most funny, some practical and some gruesome. Like the stories of the notorious Stewarts who would have had all the islanders killed off if they had their way. Thankfully, that is in the past and although not forgotten, it has probably helped to give them their resilience. That and the weather.

There's a lot of characters like Lewis the Lighthouse Keeper AKA Bubblewrap man (your secret is safe with us, Lewis). Then there's the Stonemason who introduced us to rum and coffee toddies that nearly takes a bottle of rum in one sitting, the ferryman who takes you to Mousa can tell you a few things and of course, you have the cutest little ponies you'll ever see. The otters, the orcas, the Aurora, Viking Boats, Up Helly A, the mysterious Broch and the ancient ruins which make you feel transported back in time. Combine all of these and you have just a little of the big taste of Shetland.

At this point, I would like to give a very big thank you to the kind lady for the use of the old boat storehouse in which the first half of the book got to be written in comfort and the same thank you goes to the use of the excellent facilities here at Aithsvoe Marina.

We've had a great time in Shetland and just as we were heading for the ferry, hoping for a good sailing, we get a going-away present: *hailstones*!

www.ingramcontent.com/pod-product-compliance
Lightning Source LLC
Chambersburg PA
CBHW071358170526
45165CB00001B/92